Guilty To Driza-Bone

Frank Fisher
with Lowell Tarling

crowncontent

Published by
Crown Content
A.C.N. 096 393 636
A.B.N. 37 096 393 636
75 Flinders Lane
Melbourne Vic. 3000
Telephone: (03) 9654 2800
Fax: (03) 9650 5261
Internet: www.crowncontent.com.au
Email: mail@crowncontent.com.au

The National Library of Australia
Cataloguing-in-Publication entry:

Fisher, Frank.
    Guilty to Driza-Bone.

    ISBN 1 74095 003 8

    1. Fisher, Frank. 2. Driza-Bone (Firm). 3. Country life -
    Australia. 4. Waterproof clothing - Australia. I. Tarling,
    Lowell, 1949- . II. Title.

687.1440994

Cover & Page Design: Ben Graham
Cover Photograph: Jean Marc LaRoque Photography

Printed in Australia by Griffin Press

To Pat and Stephen

A Coat Can Change Your Life

We're the sum of our ancestral parts
I come from Richard Mills
A convict assigned by Governor Bligh
To work at Richmond Hills

He wed the boss's daughter
That simple anecdote
Three generations later
Helped me design a coat

Helped me design not just a coat
Helped me design a style
From a century old tradition
By a man from the Channel Isles

Who took the storm-ripped sailcloth
From his fast windjammer boat
And - like the sailors have done for years -
Le Roy made a rainproof coat

Sometimes I sit and wonder
If my whole life has been planned
Le Roy too, must have felt this way
When he made that coat on-land

Bushrangers, farmers, squatters
And Australian pioneers
Have sought to wear his rainproof coat
For the past 100 years

Throughout the Great Depression,
Throughout the Second World War
The Driza-Bone coat and Akubra hat
Was what Australians wore

Especially horseback riders
And people on the land
The Driza-Bone is a rider's coat
Just go ask Princess Anne!

We bought the Driza-Bone Company
In 1974
It had ceased to manufacture
Just two years before

We brought back all its history
We moved up with the times
We relocated to Queensland
Where I made the classic design

Presidents, Prime Ministers, pop stars
And many famous names
Proudly wore their Driza-Bones
At the Sydney Olympic Games

And after all the years have past
The memories are still good
A coat can change your life, my friends
I always knew it could.

Thanks

I want to thank the following people:

- The lady who gave me my first remembered glass of milk with a milk arrowroot biscuit at the Children's Courts in 1928.
- The lady who allowed my sister to stay with me at the orphanage, so I was not lonely.
- The 'uncle' who gave my sister and me the Easter egg which taught me how to conserve special things.
- The man who drove the passenger car and looked after my two year-old sister and me, a four year-old boy, from Sydney to Dubbo, in central west New South Wales.
- Uncle Edgar and Auntie Violetta who gave as much as they could afford to city kids who were dumped on them during the Depression and who taught us discipline.
- Uncle Bill who showed me how to build a house from leftover timbers and who taught me to swim in the Macquarie River.
- Uncle Rowley who showed me the value of time.
- Tommy Rice, who paid me my first 3d for doing a full day's work cutting out cotton bushes.
- The doctor who operated on me for the mastoid operation, who explained how he had kept me alive and made me unafraid of death.
- The teaching nun at the Parkes Catholic School who unsuccessfully tried to teach me religion, but succeeded in teaching me the 10 Commandments which have been my guiding light throughout my life.
- Ron Millar who gave me my first job as a company representative selling industrial safety equipment.

- Clive Nicholson who taught me to set my horizon high.

- John Partridge who taught me to be a good salesman and what not to do if you want to stay in business.

- Fia, my production supervisor who gave her full loyalty and who was prepared to argue her point of view of how things were to be made. She and her husband moved from Sydney to the Queensland plant and accepted the position of caretaker in the plant's own unit. Her participation was invaluable.

- Gordon Harman for his contribution to the manuscript in allowing us to interview him and in granting us his full cooperation.

- My son Stephen for his contribution to the gigantic move from Sydney to Queensland and to the marketing and sales of Armour Safety and Driza-Bone products, wherever sales could be made all over Australia.

- I must also thank the thousands of people who worked with and for me over the years 1962-1989, in Australia and in New Zealand until 1994,

- Above all, my wife Pat who had the most significant effect on me. She taught me to give of myself, to love someone without a barrier, to have confidence in myself and to know that the bad times are easier if shared with someone who cares, and the good times are so much better if also shared. Our lives were enriched because of the love we had for each other. We were always best friends.

Contents

Introduction	**xi**
Part 1	**1**
Guilty	3
My Childhood	13
47 Odd Jobs	39
War!	49
Travellin' Around New South Wales	65
Patricia	79
Mr & Mrs - Driza-Bone Comes Later!	97
Part 2	**109**
Safety Salesman	111
Superior Safety Salesman	135
Our Own Business	151
Industrial Safety Equipment	165
Part 3	**179**
A Raincoat Is Protective Clothing	181
Two Great Men: Emilius Le Roy & Don Pickup	195
Gordon Harman And Driza-Bone	207
Making Driza-Bone Great	213
The Legend	237
Selling The Company	265
Proving A Point	281
The Best Is Yet To Come!	287

Also by Lowell Tarling

Fiction
Taylor's Troubles (1982)
The Secret Gang of Oomlau (1988)
1967, This Is It! (1990)

Anthologies
All The Best, A Selection Celebrating 25 Years of Puffins In
    Australia (1989)
Australia's Best Poetry Vol 1 (2001)
Visions From The Valley, Poetry of the Hunter Valley 1960-2000 (2001)

Non-fiction
Thank God For The Salvos, The Salvation Army In Australia
    1880-1980 (1980)
The Edges of Seventh-day Adventism (1981)
The Australian Book of Letters (1989)
17 Small Business Success Stories (1991)
Gold Beyond Your Dreams (with Heather Turland) (1998)
No More Cellulite Fast (with Violette Chevell) (1999)
Beyond Azaria (with Michael Chamberlain) (1999)
Brash Business (with Geoff Brash) (2000)
The Complete Tiny Tim Interviews (2000)
Breadwinner (with Tom O'Toole) (2000)
The Women's Club (with Diana Williams) (2000)
My Dad Thinks I Rob Banks (with Joe Sammon) (2001)
Risky Business (with Clare Loewenthal) (2001)
Secrets of the Beechworth Bakery (with Tom O'Toole &
    Matthew McLaurin) (2001)

Introduction

I have a strong belief that people are the sum total of all their ancestors, as well as the sum total of all their experiences.

Logically, therefore, each person mentioned in this book will have a pedigree that will fan out into the rest of humanity – just as mine does. And in this book mine only goes back 200 years.

I wonder what Richard Mills – my earliest relative that I can trace – knew of his ancestors?

However, this is my story, the right people, got together at the right time, with the right experiences - and together they made me; also the right person at the right time.

It is also very much my history, Pat's history, Driza-Bone's history and Australian history as I remember it.

Having said that, it is also my intention to write the most comprehensive and accurate history of the Driza-Bone coat to date. I can say this because the Fisher family (Pat, Stephen and me) were the owners of the Driza-Bone company during its boom years. We have also included the recollections of its previous owner, Gordon

Harman, whose wife Hazel was a direct descendent of T E Pearson, one of the founders.

The real story of Driza-Bone began more than 200 years ago on the southern oceans. Around the same time the defining events took place which resulted in my great-great-great grandfather being transported to the Colony of New South Wales. The characteristics that he brought out from England are in me, and I feel extremely fortunate that his genes made me, because he was a great man. His genes helped make me the person I am today. And another 30,000 Australians can say the same, for they too are all descendants of Richard Mills.

My family established themselves in western New South Wales a 100 years before I was born. In time, I too spent my formative years in that district, by which time the Driza-Bone coat had become well integrated into Australian country life.

I grew up in the Depression, and like many city children, I was farmed out to live with relatives in the country. One reason for this was because no matter how bad things got, in the country you could always get a feed. The other reason was because Mum wanted to get Shirley and me as far as possible from our father, whom she was divorcing. In those times, a divorce was a big disgrace. And so my childhood years were spent moving from one place to the next, working on the farms to earn my keep. This gave me a great background in sheep, cattle, equipment and all aspects of farm work in and around Dubbo and Peak Hill NSW during the 1930s and 40s.

When I was 20 years old, I went back to Sydney for work opportunities. I came because I had a yearning for stability, security and a home. Teaming up with Pat in 1950 marked a real turnaround in every aspect of my life – from where I lived, to a change of fortune. We are a great team.

Like everyone, in post-war Australia, we struggled and we managed to keep a roof over our heads while raising our son Stephen. The next big breakthrough was starting our own business in 1962, which was a struggle at first. During this time we did all the things small family businesses do all over the country: some weeks we didn't pay ourselves, we ploughed our profits back into the business and we battled the union. It's such a relief to have gone through all that and to come out in good shape on the other side.

Our next breakthrough came in the early 1970s. At a time when we were looking to add to our range, we saw an advertisement offering the Driza-Bone company for sale or license. We were already supplying some goods to the farming community so we decided to buy Driza-Bone rainwear to add to our product range.

Although it had been around for 70 years, we bought the company when it was on its last legs. In fact, when we took it on, it was in liquidation and had not manufactured or sold any coats for over two years. Although the company had been put up for sale, during this period it had not generated any buyer interest. Then we came along, and it proved to be a magical combination – Pat managing the financial requirements of the company, our son Stephen selling the clothing and opening accounts, and me designing and promoting the Driza-Bone coats which quickly became the Australian icon.

We owned the Driza-Bone business during the most crucial 15 years of its life. And when we sold the business and retired, we did so with pride.

I look back over my years in business wondering - was it luck? Was it timely? Were we damned clever? Or was it a little of all three.

In my final assessment, I think our success was more than mere luck, because having sold Driza-Bone we then applied the very same business, design and manufacturing principles a second time when we purchased Eidex, the New Zealand Driza-Bone licensee. Again, it was offered to us when it was about to go into liquidation. We decided to purchase it at the last possible moment - on the last minute, of the last hour, of the last day of the working week – a Friday.

I replicated the business principles that had made Driza-Bone successful in Australia, and within four years Eidex Kiwiwear was a solid New Zealand business supplying local and export markets. Again, we sold the Eidex Rainwear Company for a first-class profit.

Despite this, in the back of my mind I have never quite shaken the feeling that there must have been an element of luck somewhere in our story. And I have come to believe that any luck is the fact that I am the accumulation of all the things that happened to me, my family background, the genes, and all the things that has made me the person that I am.

* * *

But it is not only me, but also Pat and Stephen. Pat's contribution to the whole is immeasurable. She is a meticulously accurate person with figures. At any moment of the day (or night) she always knew the company's financial standing and what the money was doing. Pat understood that. She made sure that the finances were in place to do whatever jobs we needed to do.

I cannot begin to measure how much Pat contributed to our success because the things that happen backstage are what makes the front stage work. Pat was always there for Driza-Bone, just as she was always there for me.

Typical of Australian family business, our son Stephen became involved and at 21 Stephen went on the road to sell Driza-Bone and other safety equipment to anywhere sales could be made, including industry, rural and retail. Stephen would travel throughout the countryside of New South Wales, Victoria, Queensland, South Australia and West Australia for some weeks at a time. He'd pack his suitcase and go to the bush. His contribution was also exceptional.

Even though he's been out of the business since 1990 he's still known in the country as 'Stephen Fisher, the man who sold us the Driza-Bones', which is a very honourable status in those old bush towns.

We were very lucky that the three of us were specialists in our own areas and we melded together very well.

As for my role, I was the main ideas man. In the factory I was pretty good at finding innovative ways to make the work run smoother. I suspect that the years I spent working in farming and mechanical environments during my childhood gave me the practical experience and common sense that proved such a boon when I ran my own factory

I got along well with my staff. I felt that I was a pretty good manager, in my own way. Nobody hated me – that I know of! Nobody got up and walked out and said, 'I wouldn't work for you, Mr Fisher!' I've never had any problems like that. In fact, I have many Christmas and birthday cards that are signed by everybody in the factory and I value them very much.

Talking about Christmas, in the old days when we employed fewer than 100 people, Pat would buy an individual Christmas present for every staff member. She knew each person well enough to know what sort of present each would like. And that lifted company morale because our people knew that they weren't all going to be herded into a room and given a Christmas sausage, a bottle of beer, or something like that. They all got something that was individually theirs.

This sort of attention to detail results in creating very good staff relationships. So we did not merely succeed with product, but also with people, which is the most important success story of all.

* * *

It seems to me that my life can be divided into three sections, four if you include my retirement years.

1. Early Years

My early years reflect the spirit of the 20s and the 30s. I was poor, as were a lot of people in the cities and the country. This was followed by the Second World War, which injected a sense of excitement into what were pretty flat and lean years.

My cultural experience also paralleled what was going on everywhere in country Australia – we listened to the radio and went to the movies when they were the newest and most exciting things in people's lives. And then we would go back to work in the shearing sheds and in the paddocks, with renewed enthusiasm with this 'knowledge' of what the rest of the world was doing.

There were new tunes, new radio plays and of course, the news broadcasts, which turned gloomy on 3 September 1939 when the 16th Prime Minister of Australia, Robert Menzies, stated that, as Britain had declared war on Germany, "Australia too is at war. May God in his mercy and compassion grant that the world may soon be delivered from this agony." As we listened to those words, we had no idea of their implications.

In the bush, somewhere in western New South Wales, nothing changes much. Not even for a war.

What did change, however, was that we had a bit more money in our pockets. The Great Depression was over for good – there was work to be done. This feeling continued into the 1950s and even into the 1960s.

2. Family & Work

The next phase of my life was most rewarding, because at last I settled down and became a part of a family – Pat's family. Mine was too disjointed for me to ever feel a part of it.

Pat's family took me in and gave me a sense of stability, which was a great joy to me. And, of course, with stability comes the opportunity to build a home, a career and to get a toehold in the business world.

Mine was not much of a toehold; however, it was something that I could build on.

I then became a salesman and I spent most of my sales life within the safety equipment sector.

While the 50s economy was boom and bust, ours was a sector that enjoyed continuous growth. I am proud that, over the years, I was involved in many of the decision-making processes that made our factories, workplaces and building sites safer places in which to work.

Another source of satisfaction was my battles with the unions, which resulted in favourable decisions for the sector in which I worked.

3. The Real Driza-Bone Story

The third phase of my life is where I am left wondering whether we were just lucky or whether we really did something genuinely clever.

But there's an old saying that I can identify with: 'I have been very lucky in business, and the harder I worked the luckier I got'.

So in the final verdict, I guess fortune smiled on us – buying and developing the Driza-Bone company was the best choice we made when Australiana was on the verge of becoming fashionable.

On these foundations we built up a high-profile business. We gained a first-class reputation through having good business ethics. We created a financially stable operation by a policy of tight money management. And we enhanced the product through a series of improvements that have made our Driza-Bone coat the famous one, worn by princes, princesses, pop stars and politicians.

However, before we get anywhere near the Driza-Bone story, let me wind the clock back to where my story begins. It starts in London where my great-great-great grandfather on my mother's side, Richard Mills, was wrongly accused of stealing.

He was then transported to Australia on one of the convict 'hell ships'. After paying his dues, Mills turned adversity to advantage. He married the boss's daughter, and his fortunes were on the rise.

Possibly as many as five per cent of the current population between Bathurst, Orange and Dubbo are descendants of my convict ancestor, Richard Mills.

Part 1

My Early Life

1. Guilty

2. My Childhood

3. 47 Odd Jobs

4. War!

5. Travellin' Around New South Wales

6. Patricia

7. Mr and Mrs - Driza-Bone Comes Later

CHAPTER 1

Guilty

Great-Great-Great Grandparents
Richard Mills and Ann Langley

My story begins on 2 April 1799, between 9.00pm and 10.00pm, when a couple of toffs - Henry Kemp and Martha Jones - were walking from Limehouse to Stepney in East London.

Suddenly, two men leapt over a hedge. One struck Henry on the arm while the other robbed Martha. They made off with eight penny pieces, 20 halfpence and one rather nice watch.

The robbery was reported to the headborough, Gabriel Herring. While searching the area with a watchman the only people in the vicinity were my great-great-great grandfather Richard Leveret Mills (19), and his friend John Bevan (40). Herring spotted Mills and Bevan 150 yards from the scene of the crime, so he took them into custody. A search of their persons revealed only a few coins and 'nothing else that anybody could swear to'. The missing watch was not on their persons. Consequently, Mills and Bevan were allowed to walk free.

Mills grew up in the Beaumaris region on the island of Anglesey, off the north-west coast of Wales. Like me, he began his working life as a 'farmboy' on a relative's farm. In those days it was a tradition that the eldest son inherited the family farm while other sons got work elsewhere. This usually meant they had to go to London. Mills' travels led him to work as a labourer in the East London area.

As bad luck – or bad justice - would have it, Bevan and Mills were re-arrested two weeks later and indicted not only for the stolen coins but also for the bloody watch. The headborough, Herring, was not consulted about the arrest until he was summoned for their trial.

The Trial Record is in the archives of the Mitchell Library, Sydney.

According to the case for the defence there were three major irregularities in the court proceedings:

(1) the victims were told by the police officers which men they should recognise,

(2) the defendants were unrepresented, and,

(3) there was no cross-examination of the witnesses.

In his defence, like Bevan, Mills denied the charge while cheekily pointing out that the police were so disorganised that they couldn't even get the date of the robbery right - having changed it from the 2nd to the 4th and back again. (I wonder if dates have some special meaning, as my birthday is April 3rd.)

But pointing out this irregularity didn't do my forebear any good. The report closes with the stark words:

Bevan, GUILTY **Death.**

Mills, GUILTY **Death.**

And for about 30 minutes or more, Mills must have believed his life was all over at the tender age of 19.

Fortunately, his neck was saved by the testimony of a 'person of note' who vouched for Mills' integrity. After weighing in the argument, the Judge commuted the death sentence to transportation. This meant exile to the Colony of New South Wales, where my ancestor was condemned to a 13-year sentence.

Mills was the first member of my family to come to Australia. Had they hung him I would not be here today, my son Stephen would not have been born, and the Driza-Bone coat would probably not exist.

Forty-three transport ships, nicknamed 'hell ships', sailed from England and Ireland between 1787 and 1801, carrying 7486 prisoners, of which 756 did not survive – that's one in 10. The suffering aboard the convict ships in the early years of transportation was appalling. Many died in their chains below deck. The sheer sadism of many of the ships' masters was legendary. The food was disgusting and the convicts were riddled with disease.

The *Royal Admiral* left England in May 1800 and after five months in a typhus-ridden hell ship, Mills and Bevan reached Sydney in November. Forty-three prisoners died on that trip, the first being the ship's doctor after only nine days at sea. Many of the survivors never regained their strength and spent years in hospital. You needed a robust constitution to survive that voyage.

Disease was not the only problem; there was also talk of mutiny on my great-great-great grandfather's very ship. I don't know what part he had to play in that, but history tells us that all the convicts were locked in double irons in the hold, until the crisis blew over.

The next drama occurred when the fleet engaged two French frigates in battle. Because the English won, the French prisoners were added to the already over-crowded, contaminated holds. And on top of everything else, Richard Mills had to cope with that. Staying alive was a full-time effort.

Governor King was horrified at the condition in which the prisoners arrived. Even after two years had passed, he wrote that he doubted these convicts would ever regain normal strength, such were the pains of the journey. In an attempt to reduce mortalities, in 1802 King decided to pay surgeons 10/6 for each convict landed alive in New South Wales. In Sydney, Mills was immediately sent to the colony's hospital for treatment, where he remained for several months.

Mills' first six years in New South Wales are unrecorded. His name first shows up in 1806 listing him as indentured to David Langley, a free settler, and his wife Nancy (called Ann).

The Langleys and their three daughters had settled at Mulgrave Place, Richmond Hill (now known as Richmond, 8 km away from Windsor, NSW). Two years earlier, the Langley family had received a 100-acre grant from Governor King, and their home 'Clear Oaks' still stands today, in excellent condition.

Being a skilled artisan from London rather than an experienced farmer, Langley probably delegated all the important farming duties to the convict Mills. When, in 1806, the new 'Governor' William Bligh appointed Langley to the post of Superintendent of the Government Lumber Yard in Sydney, this set the Langley family up with a town residence and a good degree of security. And again Mills was his anchorman, keeping the home fires burning.

At 16, Langley's daughter Ann fell for him and by the end of September 1806, she was pregnant with Richard Mills' child. Their daughter, Harriet – my great-great grandmother - was born in June

1807 and registered in Mills' name. The couple formally married four years later.

When Governor Bligh clashed with the NSW Army Corps over his attempt to control the rum traffic, Langley's loyalty to Bligh cost him his post at the lumberyard. Harder times followed, until Governor Macquarie eventually reinstated Langley, and life in Sydney and Richmond Hill continued as before: comfortable for Langley, hard-working for Mills.

Richard and Ann married on 11 June 1811 and four months later, their second child, Richard Langley Mills, was born. A year later, in December, came Elizabeth. And the following month, on 31 January 1813, Mills became a free man at last. In 1814, he is listed as a free landholder.

Ann and Richard prospered; they had 11 children: six boys and five girls.

In 1815 Governor Macquarie chose the site for Bathurst, which triggered the expansion and development of western New South Wales. The city of Bathurst is the first established settlement west of the Great Dividing Range. Also in this year Cox's Road (connecting Emu Plains to Bathurst) was opened. Mills would travel this road two years later when making five-day journeys - carting supplies beyond Mount Victoria in the Blue Mountains region to the Cox's River area, south of Lithgow and Hartley. By this time Mills was an enterprising married man. He had a horse and cart of his own and he earned 12 pounds 10 shillings per trip.

In 1818 my great-great-great grandfather applied for a grant of land in Bathurst. The fifth name on the new settlers' list in Governor Macquarie's diary, initialled 23 April 1818, records the new land-owner as 'Richard Mills (conditional pardon)'.

Richard, Ann and their five children established their new home in Bathurst where they built a wattle-and-daub cottage.

While Ann was setting up her modest household, Richard was at work on his 50 acres of farmland. Like the other first nine Bathurst settlers, they had also been given two acres in town, a cow, food supplies, four bushels of seed wheat and one convict servant to labour for them, for a period of one year. The city of Bathurst enjoys moderate weather throughout the year, which favoured Richard and Ann's wheat crops.

After this, Richard and Ann moved into material security. Unlike most of their neighbours, they did not raise sheep; they bred horses instead.

As their fortunes as well as their family continued to grow, they enlarged their home. In time, they opened the King William Inn, the third licensed inn on the Bathurst Plains.

The Mills family and their descendants spread way out into western New South Wales, as far afield as Mudgee, and probably further. Their most consistent occupation was running farms or taverns.

In time, the patriarch Richard Mills, his wife and as yet unmarried children, moved to Mudgee. There they lived with their daughter, Mary Ann, who was married to a prosperous hotelier William Ealy Sampson. Sampson had owned a pub in Sydney, before heading west and building the Mudgee Tavern. Since his continuing Sydney business interests required him to travel a lot, Richard and Ann managed the tavern during his absences.

Richard Mills died in Mudgee on 12 December 1850 at 69 years of age. His body was carried over the Blue Mountains and buried in the Holy Trinity churchyard at Kelso.

My great-great-great grandmother Ann outlived him by 25 years.

Great-Great Grandparents
William Young and Harriet Mills

On 7 December 1825 Richard and Ann's eldest daughter Harriet – my great-great grandmother - married William Young in St John's Cathedral in Church Street, Parramatta. Four days later, Harriet's younger sister Nancy was baptised in the same cathedral.

It was a big December. When Christmas came two weeks later, the family had much to celebrate.

Harriet and William lived many years in Kelso, in the Bathurst area, on a portion of the farm given to them by Harriet's father, Richard Mills.

In 1846, after her parents had moved to Mudgee, Harriet and William moved to a station called The Grove, at Trunkey Creek south of Bathurst, in a wide valley of the Great Dividing Range.

Like her mother, Harriet also had 11 children.

Great Grandparents
Josiah Young and Isabella Paterson

Josiah Thomas Horatio Young, was born at The Grove just two years before Richard – the patriarch – died. My great-grandfather was the ninth child of Harriet (Mills) and William Young.

Josiah married Isabella Paterson, who bore him 10 children. Their fourth child, Frances, was my grandmother.

Interestingly, Australia's most famous poet, Andrew Barton 'Banjo' Paterson, has the same surname, spelt the same way. He was born at

Narambla Station, near Orange, which is in the same region as my folks.

Another link with history was that one of Harriet's sisters, Mary Jane Mills, was visited by the bushranger Captain Midnight (Thomas Law) during 1877-1878 at her Wondobbie Inn on the south bank of the Marthaguay River. When Captain Midnight was shot down at Bourke on 3 October 1878, Mary Jane's handkerchief was found on his person.

Grandparents
Clarence Curran and Frances Young

My grandmother, Frances Morano Young was born at Trunkey Creek near Bathurst in 1881. She married Clarence Wellington Curran of Wauchope, northern NSW. They had three children, each of whom played a significant part in my early upbringing. They were:

- Clarice (Claire) Frances - my mother,
- Lila Morano (Auntie Lila), and,
- Farrell ('Foul House' or 'Uncle Bill') Wellington.

My grandfather Clarence was a wharfie by trade, specifically a 'trolley man'. Frances and 'Clarrie' lived most of their lives in Sydney, near the wharves at Woolloomooloo and later in Paddington.

He was a heavy drinker; in fact, I had never seen him sober. In those days each trolley man who worked the wharves carried bags at the slowest speed so as to get more money for the work. They were also entitled to one 'free' trolley of goods at certain regular times, in turn. This contraband was then sold and the proceeds pocketed. The practice was well known and accepted on the Sydney wharves at the time.

Having been born four months before her parents married, my mother – Claire – might have felt she was not a wanted child.

Perhaps she felt that her father had unwillingly married her mother, which could account for the 'attitude' my Mum always bore.

I am proud of my maternal line – Mills, Langley, Young, Paterson and Curran. As a pioneering family in western New South Wales, all these people and their offspring played a big part in the settlement of that part of Australia.

In fact, the original Langley homestead at Richmond, NSW, is still lived in today, not by my relatives, but by the Livingstone family, descendants of the Scottish explorer and missionary, Dr David Livingstone.

Conclusion

Looking back, I am amazed that the decisive moment came nearly 200 years ago in 1818, when Richard Mills traversed the recently opened Blue Mountains Road and – with eight other families – settled an area which has become Bathurst, Paterson and Trunkey Creek.

Many members of my direct and indirect family still live in those parts today.

CHAPTER 2

My Childhood

Bette Davis has got big eyes and she's always so fierce. I don't like Bette Davis movies because she looks like Mum and reminds me of her darker moods. Everything was a big deal. Mum was always heavy like that.

I didn't live with my mother much. She was always vaguely in the background of my life – seldom in the foreground.

Mum was born in Sydney on 18 December 1906. She was five foot three in height, slim, with dark hair and brown eyes. But my mother was not a happy person. She was cranky, always cranky; all my life she was cranky. She was always sick. Always complaining. Never happy, never content. She eventually died from a cerebral hemorrhage when she was 42. She must have been sick for donkey's years, and that's probably what made her so cranky. My mother was forcing herself to stay alive.

Mum was a very good cook, though by trade she was a machinist, She was machining when Surry Hills was the centre of the rag trade in Sydney, around Central Railway Station. Perhaps my lifelong affinity with sewing machines stems from my mother.

When I was little, Mum visited a place in Alice Street, Newtown. I remember Alice Street very well because there is a park across the road from a row of terrace houses. My mother used to visit somebody who lived in one of those terraces. I sometimes accompanied her and she would leave me to play in the Matt Hogan Reserve playground opposite, while she went off to see friends.

I recall one particular incident very clearly because I had long hair at the time. Mum heard another mother tell her child, "Go and play in the playground with that little girl."

My mother was most upset. She whisked me inside and immediately chopped off my hair, from long, girlish curls to a very short haircut.

I went back to Alice Street not so long ago, and memories like these all came flooding back.

Nursing Home In Kent

As for my father, Gilbert Edmund Fisher – his parents were doctors who were superintendent s of a nursing home at Sandgate, on the Kentish coast, south of Canterbury near Folkestone, England.

One story tells that his parents died in a car accident; another tells that they died in a fire that destroyed their home. Either way, they were tragically killed. I tend to think that they were burned to death, as I have been told stories of the destruction of the nursing home by fire around 1921/22.

A Comfortable Past

My father was born on 4 October 1903 and lived a comfortable life until the untimely death of his parents changed his fortunes. He was probably shunted from pillar to post until he made his way to

Australia where he was housed in a home for destitute boys. I assume he graduated from there as soon as he got a job. And then he met my 16-year old Mum and started dating her. A whirlwind romance led to their ill-fated union.

Physically, my father was 5 ft 10 in tall and weighed about 13 stone. When I knew him, he was quite bald and wore glasses because he was shortsighted. He struck me as being a 'gentleman'. He had obviously enjoyed the privileges of a very good education because he had excellent command of the English language.

However, in reality, my father was a con man. He probably felt that life had somehow cheated him, which gave him the right to cheat others. My father never achieved any financial significance from the time he set foot on Australian soil until he died.

He married my mother on 30 January 1925. They had two children, me – Frank Gilbert, born on 3 April 1926 – and Shirley Lila, who was born 20 months later.

My First Years

I was born at 3.30am in the Royal North Shore Hospital in Sydney. My mother had a rough birth, and I did not come out in top condition but I survived. Here I am today.

At that time my parents lived near the railway line in Alfred Street North Sydney, though the address on my birth certificate is the suburb of Willoughby. Therefore, I am uncertain where my first home was, except that it was in either of those two places on Sydney's lower north shore. The little I remember of my parents' lives together amounts to nothing more than living in cold and bedbug-ridden flats.

My parents didn't seem to live in any single place for any significant length of time. They were always on the move, ultimately moving right away from each other. From the time my mother ran away from my father, my life was in the lap of the gods.

I was introduced to some 'uncles' especially during that period of time. Uncle Emmanuel called around more than most. At Easter 1930 he gave Shirley and me an Easter Egg that was as big as an emu's egg! We were so thrilled because we hoped it was solid. However, it was made out of light and very thin chocolate, of course, and the egg had nothing inside. But it was much better than nothing. We made our Easter eggs last for days and days, because we weren't getting regular meals at that time and we had to conserve food if we wanted a treat for the next day.

One of the highlights of this part of my life was when I fixed a little Dinky tricycle. It doesn't seem much, but when you've got a three-wheel trike and you don't have any pedals on it, there's only one way it can go - that's backwards into walls. So I got two bolts and made two pedals so that I could go forwards as well as backwards. I got great pride out of that.

It probably inspired me to go on fixing things over the years.

But the story didn't have a happy ending because now that it was fixed it was worth something at a time when money was scarce. So my Dinky trike was sold – boy, that hurt!

The Fall

One of my earliest memories is seeing my mother lying at the bottom of the staircase. At the time I was three years old.

We lived upstairs and I was in the hallway. People had gathered around trying to figure out whether my mother had been pushed or whether she had accidentally fallen down those stairs.

After this a magistrate decided that my mother could not look after us adequately. Shirley and I were taken to the Albion Street Children's Court in Surry Hills where we were declared Wards of the State.

Around the same time, my father filed for divorce. My parents separated in 1928 and the divorce was granted in 1936. According to my father's side of the story, he was granted access rights to us. But at her first opportunity my mother whisked us away to the west and my father didn't find us for years. When we found him, he told us that he had spent quite some time trying to find us, but had not succeeded. I am not convinced that he tried very hard. Given her family background, where else would Mum take us?

But that was the future; my immediate problems were more pressing. Scared and hungry while sitting in the Children's Court, a kind lady came up to me and gave me a full glass of milk and two milk arrowroot biscuits. I cannot begin to express my gratitude at this gesture. Even now, when I eat an arrowroot biscuit, a profound sadness comes over me as I recall what it felt like to be taken from my mother and sent to a children's shelter.

Shirley and I were sent to live in a refuge in Forest Lodge, a tiny suburb between Annandale and Glebe. (Shirley and I have since returned to Forest Lodge. We looked around and we reckon that we have found the old place.) To us it was an 'orphanage', but it was a 'refuge' to others, as there were some intellectually disabled people living on the premises too. There were a number of people in that condition in the various places where we stayed.

I saw my first dead person while staying at that orphanage,

We kids used to spend a lot of time playing in the streets and in the back lane ways. One day we were standing across the road from a man who was leaning over his car engine. For some reason, Shirley was sitting in the car while the man was pouring petrol into the carburetor. There was a sudden flash. The explosion nearly blew his head off his shoulders.

The infusion of petrol shot the car across the road where we children were standing. It smashed into a gate, narrowly missing me, though it grazed a couple of the other kids.

Miraculously, Shirley was unhurt.

Orphanage Days

I suppose some people were in charge of the orphanage, but we never saw them; we just hit Bridge Road and mucked around. I was only three at the time and too young to attend the local Forest Lodge Primary School. We just stood around and watched whatever was happening in the streets. And, of course, like about a third of the population, we were always hungry.

They only fed us one meal a day at that orphanage – a large helping of rolled oats porridge served in a bowl at breakfast. We were also each given a slice of toast that was dipped in Cocky's Joy – Golden Syrup. The Cocky's Joy was watered down and heated in a pot, after which the cook dipped the bread in, then let it dry out on a big tray before bringing it out for us to eat. It tasted beautiful! I still love Golden Syrup.

In 1930 the Great Depression came. The national income declined from £640 million to £560 million. Unemployment increased sharply, and within a short time nearly a third of all breadwinners were unemployed. We sometimes stood around and watched the unemployed people clamour for work or relief. Soup Kitchens were prevalent all over Sydney. At lunchtime we used to join the queue to get our helping of soup in our little mugs. If we got there too early, all we got was the water on the top.

After a time we became smart and realised that if we came in last, or nearly last, we would get all the ingredients that had sunk to the bottom of the pot. But sometimes we'd miss out altogether because if you were too smart and arrived too late, you got nothing. When the pot was empty, that was the end of the soup.

The evening meal, if we got one, was provided by an 'uncle', Mum or somebody like that. Mum would come to see us semi-frequently, very often bringing us delicious custard. She would sometimes take us to play in a park or maybe as a special trip she would take us to the beach at Manly.

Going West

After a while we left the orphanage and moved in with various relatives and friends. First we lived somewhere in Paddington, then in Darling Street, Glebe, and then in Bridge Road, Glebe. It was a real mix-up; something permanent had to be arranged.

I was four when Shirley and I went west, to live on Auntie Violetta and Uncle Edgar's farm in Eumungerie. Violetta Young Hackney was my grandmother's sister (that is, Frances Young's sister). We called her 'Big Auntie' because she was a big lady with a big presence.

Mum had two reasons for sending us out west: the first was because it was easier for people to find food in the country than in the city during the Depression. The second was so that my father could not find us. This was a pretty miserable thing to inflict on two little kids who would spend more than a decade wondering about their father.

Going west was a saga all of its own. Shirley and I were transported in a type of mini-bus, simply called a 'big car' in those days. It was big because drivers would extend their vehicles to fit in as many passengers as possible. These were stretched Buicks, Packards – even Rolls Royces! These 'big cars' used to surreptitiously pick up passengers from Martin Place in the heart of Sydney, and take them to country destinations.

It was an open car with a canvas top, wooden seats and two foldout chairs behind the driver's seat. The nine passengers were bumping and bouncing all over the Blue Mountains on our way to Dubbo. This 1930s mini-bus offered nothing remotely resembling luxury. In fact, its driver had his hands full not only dodging potholes but also trying to duck the watchful eyes of the inspectors riding the trains to spot such illegal conveyances.

In the 30s it cost £3 to travel from Sydney to Dubbo by rail, whereas by commercial transport it only cost £2. So 'big car travellers' would save £1 on the trip, which was a significant saving when money was tight. However, it was illegal to provide such transport alternatives, because this competed with the railways.

All the cars leaving from Martin Place were supposedly going no further than Parramatta or some outlying Sydney suburb. In fact, they were going to different parts of the State, with passengers getting on and off like bus travel.

All of a sudden, in the middle of nowhere, the car came to a halt.

"All right, ladies to the right and gentlemen to the left!"

Out scrambled the passengers to avail themselves of this unceremonious comfort stop behind the thick growth at the sides of the road.

"Go on Shirley," I said, "Off to the right."

No adult was accompanying us, but we were used to that. After the confines of the children's shelter, this trip was a real adventure.

Finally, Shirley and I arrived in Dubbo. Mum's brother, Farrell ('Foul House') Wellington Curran – better known as Uncle Bill – and his wife Auntie Mae, met us and housed us for the few days until Big Auntie Violetta came into town, to pick us up and take us to her place.

Dubbo is on the Macquarie River. In flood the water moves very slowly, which shows how flat it is. Flood water just sits around for days and sometimes many weeks. The worst part of town was the flats, which is where Uncle Bill and Auntie Mae lived.

Uncle Bill was unemployed. He and Auntie Mae lived on the Dubbo Flats with all the hoboes. Shirley and I moved in with them for a few days, awaiting the arrival of Big Auntie.

Although Shirley and I didn't know it at the time, driving west from Sydney was a trip backwards through our family history. We travelled from Emu Plains over the Blue Mountains, to Bathurst, Orange, Narromine, across the Harvey Ranges, into Dubbo, and ultimately Eumungerie and Peak Hill – where the Currans, Patersons, Youngs, Langleys and Mills' had forged their place in regional Australian history.

(More than 13 years and 40 jobs later, I would come back over those Blue Mountains on my way to Mum's home in Sydney. This time I would be riding my trusty 1927 Harley Davidson motor bike.)

Eumungerie

Big Auntie came to town twice a week with cream – I don't know how she kept it fresh – and she met us, picked us up and took us to her farm in Eumungerie, a town halfway between Gilgandra and Dubbo.

You would not even slow down in the main road of Eumungerie, yet at one stage it had a population of more than 1000. The population of Eumungerie might now number 20 but it has a strange history because at one stage it was an important part of the national tennis circuit. Of all things, a number of tennis courts were built there as the Outback's answer to Wimbledon. Their ghostly shells still stand.

The wide and dusty road to Eumungerie crossed the Drillwarrina Creek. The railway line lead to the township which consisted of not much more than a church, a store, a one-room schoolhouse, a pub and a playing field. The countryside is mostly flat and brown, with clusters of Cypress Pines, Peppercorn, Red Gums, Ironbark and Kurrajong trees – that typical parched outback look.

The Hackney farmhouse is still standing, on Goonoo Street near the corner of Eura Street. There were shearing sheds out the back as well as a derelict shearer's hut, which is where we kids used to sleep. It was freezing in winter and sweltering in summer.

In the days when Big Auntie and Uncle Edgar housed us at Eumungerie, up to 10 other children of relatives lived on the premises, all dumped on them to look after, so they had to be very strict with us, I guess.

For example, meal times were a real ritual. For starters, we all had to sit in silence on the long bench after which Uncle Edgar made his entrance. We kids all sat along the wall and the adults sat around the other parts of the big table with eight, 12, sometimes even 14 people all expecting to be fed. In those days kids were to be 'seen but not heard'. It was a long table and we sat there with long faces.

We kids were all served second, the adults were fed first and what was left over was divvied up between us. If there were 10 kids and only enough food for four, we got whatever that percentage was.

Furthermore, we weren't allowed to talk at all, not even to ask for salt, butter or a slice of bread. If the bread was way down the other end of the table and one of us wanted a slice, the only way to get one was to stare at somebody at that end and hope that person would pick up the vibes and say, "Would you care for a piece of bread Frank?" "Yes, please," I'd say, and only then would they pass me a slice.

Uncle Edgar had an order of preference regarding our sitting positions. Before the meals were served, we had to sit with arms *folded*. While we waited for our sweets, we were supposed to sit with our *hands on our laps*.

We were never allowed to put our elbows on the table. We were never allowed to use a fork upside down. We couldn't just dump our knife and fork on the plate. We had to put our knife and fork in a dead straight line with the cutting edge pointing towards the plate. Even today when I put a knife down on the table, I still position it with the cutting edge towards the plate. I can't help myself. It's a habit I cannot break.

We were also taught to eat peas from the top of the fork. For some reason, in my four-year-old mind, I felt that Uncle Edgar's credibility largely rested on these dinner table rituals. Much later in life I came to see for myself that not even the Governor of New South Wales ate his peas that way.

I suppose being extra-strict was the only way Uncle Edgar and Big Auntie could control the place, because they always had kids turning up for a few days here and a few days there. If they hadn't imposed some sort of system, a situation like that would have been bedlam, because the strict regime didn't stop us putting snakes and spiders in Uncle and Auntie's bed at night. In retaliation, Uncle Edgar imposed even stricter discipline. We were generally punished with a hit from a strap on the hand or on the backside. He and Auntie also had other ways to make sure we obeyed the rules, which was deprivation – and worst of all – no pudding!

We didn't complain about it; we just accepted that we were always hungry. The only time I remember being 'full' was at the annual Picnic Race Day.

Picnic Race Days are something you look forward to in country towns. They have cake stalls, craft exhibits, a sausage sizzle and meal tents. I especially remember my first Picnic Day; all the food was free – baked dinners and everything! I can remember it as if it were yesterday because I was full for the first time in my life. And I was only five years old.

The Picnic Race Day would always generate great excitement throughout Eumungerie. The women would get to work making their crafts and exhibits, and the men would work on the horse races, car races, utility races, kids races – the men were always race-mad at the Annual Eumungerie Picnic Day. It was marvellous.

Kids' Farm

Big Auntie and Uncle Edgar's 'Kids Farm' wasn't pure charity. We had to earn our keep. In the days of the Great Depression not too many people could afford charity.

I was assigned my first job on the very first day I arrived, and I mucked it up terribly. There were 10 children living there at that time and I was the eldest boy. So Uncle Edgar turned to me and sternly handed me a white aluminium container and said, "Go get some chips."

What I hadn't realised was that he wanted me to go to the wood heap, pick the wood chips from the ground and return when I had filled the aluminium container from which he would start the fire.

But I had only just arrived from the city. As far as I knew chips were made out of potatoes. I wandered around for ages before returning very distressed because I couldn't find the fish and chips shop. What a crummy start!

I had been really looking forward to eating those potato chips.

Instead, I was told in no uncertain terms that I had not done my job properly.

However, after I had been at Eumungerie for a while, things settled into a routine. My daily roster was to get up at daybreak, fill the hot water bucket, fill the firewood box, let the dogs off their chains, bring the cows in, then wake the other kids and bring them to the farmhouse for breakfast.

Then, because I've got fairly strong hands, I was quickly assigned to the dairy – and I learned to be a fairly good milker. We milked a herd of 20-30 cows at that time – twice a day, of course.

Another of my jobs was to keep all the kids out of Uncle Edgar's 'hair' when he was working. He made this order even though he was nearly bald. We kids made up lots of cheeky stories and jokes about 'being bald' when we were out of his hearing.

That didn't leave me much time for inventive thought. However, I had noticed that there was a creek on the property, an offshoot of the Talbragar River. And it was full of carp.

I made a fish trap out of a four-gallon kerosene drum and figured out how to catch them in the shallows. Because I had only managed to trap tiny fish, I didn't bother offering them to the cook. But where should I keep them?

I put them in the horse trough.

The fish frightened the horses and Uncle Edgar gave me a strapping for doing that.

As I got better at trapping fish, I started to catch a few that were big enough to eat. Big Auntie appreciated that; she cooked them for supper and divided them amongst the kids. Unfortunately, one of her granddaughters got a fishbone stuck in her throat and it nearly choked her. She made a helluva fuss. It was a right carry-on, which put a damper on my fishing endeavours – for a time, at least.

Eumungerie Primary

Being only five, I was probably the youngest student at the Eumungerie Primary School. The total attendance varied between eight to 10 kids made up of six classes in the one room.

I was quite thrilled to attend school because I wanted to learn to read and write.

I remember the first time we were able to draw on a slate with a slate pencil. I remember that tiny one-teacher, rust-red weatherboard school standing beside the railway line complete with verandah and tin roof.

Guns and Horses

Auntie Violetta and Uncle Edgar owned two cars that they seldom used – a Dodge and a Rugby – and both cars fascinated me. I was so disappointed that they always used the horse and sulky instead. (They simply couldn't afford the petrol, but to a child's mind it seemed like a real missed opportunity.)

I was in the bush when I first rode a horse. Farm kids don't remember the first time they rode a horse as being a special day – we just did it. If we had to go somewhere, we would jump on the horse and if we got thrown off we'd try again until we could ride.

I also had a 22 Remington rifle, but I wasn't much good at shooting because I was shortsighted, though out of necessity I've shot most animals on the farm. I didn't know I was shortsighted until I tried to join the New South Wales Police Force, years later.

I remember shooting a rabbit up the backside. The bullet came out through its mouth and there was no mark on it, but it was dead – and its skin was worth twice the price of a skin with a hole in it.

That Old Pianola

The only music we knew was played on Auntie Violetta and Uncle Edgar's pianola, which I used to occasionally play.

Playing the pianola was always a major event, we didn't have a bit of a twiddle and then leave. It wasn't something we did casually, it was a 'big deal' because the pianola was the only music we heard. We prepared ourselves for the occasion, everybody sat around, twittered and got excited about it.

When the time came for anyone to play the pianola, everybody knew about it and prepared for it. Everybody got involved and enjoyed it, and some would even sing along.

But pianolas are not a versatile instrument to sing with because they are in a set key. So our musical evenings had their limitations – unlike the rest of the farm, which was boundless.

School Of The Air

In 1931 Mum left Sydney and came west too. She found work around Dubbo and Wellington, so Shirley and I left Eumungerie school, joined up with Mum and did correspondence school, and sometimes School of the Air. Shirley was allowed to accompany Mum to work, but I would be sent away somewhere else, because I was a boy.

We were living at some station near Narromine when Mum enrolled us into correspondence school. I remember her sitting with me and making me study things that I didn't want to learn. I was okay at Maths but she forced me to do History, which was all about Kings, the Norman Conquest, the Battle of Hastings and other aspects of English history.

The history of Australia was absent. Even though we lived here, Australia didn't exist as a place as far as the History Section of the Department of Education was concerned.

As a divorcee, Mum was not an unassuming figure. Women criticised her and men 'noticed' her everywhere she went, including one of my cousins, which did not help. But Mum came into my life; she went, came back again and went off again – all in the course of her work. As I remember her, Mum was always 'somewhere else'.

I was not at Eumungerie for long before Big Auntie would pack me off to stay with other relatives. I travelled back-and-forth between Dubbo, Tomingley, Eumungerie and Peak Hill – and a couple of times up to Wauchope on the north coast of NSW where I would stay with the Currans or the Kellys.

Consequently, I was dumped into Big Auntie's care quite a number of times – or Uncle Bill's, or Uncle Rowley's – sometimes to anyone who would take me. So too were my cousins, Jeannie and Joycie Collison, the Tremaine girl and a couple of other boys whose names I cannot remember, and you can add the Hackney children, to the numbers having to be housed and fed. We were all farmed out to wherever we could get a bed, and neither Shirley nor I was permanently located at any stage.

Tomingley was another place we called home where we stayed with Big Auntie and Uncle Edgar's son, Gordon Hackney. Gordon was the stock and station agent in that little village on the Newell Highway between Dubbo and Peak Hill.

Today it calls itself the 'Gateway' to the Narromine Shire because that's precisely what it had been when it had 40,000 inhabitants before the mines petered out.

Sydney Harbour Bridge

And then Mum took us to Sydney for a brief stay in Forest Lodge, our old stamping grounds. We celebrated the opening of the Sydney Harbour Bridge in March 1932. Shirley was in the first group of children to cross the Bridge. I was in the second. One month later Phar Lap, the famous racehorse, died.

Then Shirley and I went back west, this time by train, travelling in the last car – the guard's van.

By 1932 the number of jobless in Australia had grown to 337,000 and Australia did not begin to recover from the Great Depression until 1939, but I never felt poor. Everyone was the same.

Sometimes we ran into swaggies travelling from town to town. They weren't dangerous. They all had their swags hanging off their shoulders, their billies and a bluey cattle dog walking beside them in their travels around the countryside. We thought of them as ordinary blokes walking around trying to find a feed.

Because they had travelled, the swaggies were our main source of news in those days. Sometimes Big Auntie would give them a feed because they could tell us the news from the outside world or what was happening in whatever town they had just come from. They were all very nice people. I can't remember one man – or woman – among them who was bad.

I say 'women' because people forget that there were women swaggies too. I don't ever remember them being particularly badly dressed, but they all seemed to be comfortably covered. They all wore great coats, overcoats or leftover coats from the First World War. A lot of them had made their own coats. Sometimes they would use wheat bags over their shoulders like a cape. Some wore flour bag clothing, because in those days flour used to come in calico bags with which we used to make shirts.

Around this time I first became aware of the windjammer coat.

Because they were so distinctive, I noticed them on the farms when Shirley and I were sent back to Dubbo.

Back and Forth

Next, Big Auntie and Uncle Edgar had moved from their Eumungerie farm to run a mixed goods shop in Dubbo so we were back to Uncle Bill 'Fowl House' Curran's place on the Dubbo Flats. First, we lived in a disused movie theatre, then in a shack by the Macquarie River, in an area known as the 'humpy flats' or 'tent town' where unemployed people like Uncle Bill lived.

Aboriginal and poor white families lived side by side in hessian bag humpies – although Uncle Bill and Auntie Mae reckoned they lived in 'luxury', their hut being built out of bark and proper logs. Uncle Bill showed me how to build a two-bedroom hut out of leftover timber from the timber yard.

Then the enlightened Mayor of Dubbo, Dr Gordon Fitz Hill, hit upon the idea of 'working our way out of the Depression'. Fitz Hill got the unemployed population to work on developing Victoria Park, setting flowerbeds, planting lawns and digging the Olympic Pool. As far as my health was concerned, it was a pity that pool hadn't been built much earlier because I learned to swim in the Macquarie River instead.

Along with two other boys, I caught an infection from swimming in the dirty and polluted river water. We were operated on for a mastoid infection. I lived; the other two boys who had the same problem, died.

Doing Whatever Needed Doing

Going from one farm to another and being the boy around the place, I had to do what had to be done because I was the only person available to do it, or sometimes because I was the only person who was prepared to do it.

This attitude has stood me in great stead over the years. I have made a success of my life because of what I experienced in those early days; there is no doubt about that. It probably was tough, but it made me very adaptable.

I did what the job required; I never thought of myself as an 'adult' or a 'kid'. I couldn't say 'I'll only do a child's job'. Like everybody else, if there was a job to be done, and I was capable of doing it, I'd go ahead and do it. We didn't say that's 'a man's job' or a 'woman's job'. These days there's a line of delineation; it didn't happen in the 30s. There were no division of labour on the farms between a man's job and a woman's job.

If a woman was capable of driving a team of horses, she drove a team of horses. If a man wanted a meal, he went and cooked a meal. What had to be done, had to be done.

Time Management

My next memories are of staying with various members of the family at different times. For example, Auntie Violetta's brother – Uncle Rowley – who taught me the value of time - lived 70km southwest of Dubbo at Peak Hill. I lived with him for a while. He owned a dairy farm and was a supplier of milk to Peak Hill. Compared to Eumungerie, this was a big town. It had three hotels.

Peak Hill had two milkmen and Uncle Rowley was one. In those days people would leave a billy can at the front door with money inside to show how much they wanted. (Example: 3d a pint, 6d a quart and so on.)

Time is the most valuable thing there is. The milkman who provided the milk was the one who arrived first. Uncle Rowley, his sons Hector and young Rowley ('Cobby') and me, would get up at 3.00am, race out to the dairy and milk the 30 or 40 cows, then rush into town and start delivering the milk before the other milkman could get there.

It was a real contest because milk deliveries gave the best return. If we were late, we had to separate the cream and try to sell that, but most people – except maybe the café or the hotel – could not afford cream.

So the next step was to turn the cream into butter.

Again butter was hard to sell, so it paid to get up at 3.00am and be there first.

The most important thing is *time*, and I learned that from Uncle Rowley and have applied it throughout my life.

Waterford Station

Like Dubbo and Tomingley, Peak Hill was once an important gold mining area, with a population of 30,000-40,000 at one stage. It is called Peak Hill because there was a slight peak there once. Now it's the very opposite of a hill; there's an excavation where it was. There were a lot of mines, emus, kangaroos and wild pigs in Peak Hill. And everything was flat, parched and dry.

I grew up with Aborigines in Peak Hill, but we did not think of them as being different. They were hungry, and we were hungry.

I remember two Aboriginal boys called the Solomon brothers. One of them became the first Aboriginal managing director of the Aboriginal Board.

Eventually, Mum got a job earning 20 shillings a week as a housekeeper, at a merino station called Waterford Station six miles out of Peak Hill, where we subsequently lived, mostly together, for the next two years of my life.

The station was owned by Tommy Rice, whose wife and son had died, and he needed to employ a housekeeper. Mum took the job. Tommy Rice decided that Shirley could move in too, but there was no way in the world he was going to feed a growing boy, because I had hollow legs.

But Tommy got over that and he eventually agreed to let me live there too. So, in 1935, we moved to Waterford Station and Shirley and I attended the Peak Hill Primary School.

At first we walked the six miles to school, then Mum bought us a horse, ironically named Carbine after a famous 1890 Melbourne Cup winner, who was as quick as a flash. Carbine was so slow that he couldn't get out of his own shadow. It would take him one whole hour to travel the distance to school. I could walk much quicker, but riding was better than walking.

Although we didn't wear school uniform, I wore a coat and tie and we travelled to school in a horse and gig (sulky). We didn't have overcoats, raincoats or windjammer coats. To keep warm we would wrap a couple of wheat bags around us, held together with a wool bale hook, around our waist.

We used all our available resources. The flour in those days used to come in calico bags and as I mentioned earlier they were quite valuable because people would make them into clothes, particularly underclothes. The material was quite good. The women would cut them up and sew them into knickers, shirts, blouses or handkerchiefs.

School was okay in 1936. My teacher, Mary Woodhouse, was nice enough. I shared top grades in third class with a girl with the family name 'Hoy'. The Hoys are a very prominent family in the region. Although I can't remember her given name, her brother Bruce became a best friend.

I also enjoyed playing rugby league at lunchtime, but I couldn't play after school. I had to be back at the Waterford Station by 4.30 because I had to pick up the ice every afternoon on the way home and do many jobs such as putting the dogs on their runs, feeding the horses and the cows, and watering the garden.

Plus Mum would worry if we were late home. One day she had genuine cause to worry because that sulky was plain dangerous. It didn't have a proper seat and there were lots of other things that made it unfit for adults to ride – let alone a couple of tiny kids! However, the slow-moving horse made everything seem safe, until one day, Carbine shied and bolted. I tried to pull him up but I couldn't. We crashed into a gully; Shirley fell out, her head landing between the spokes of the wheel.

I just managed to pull her out with seconds to spare, before Carbine bolted again. I saved her life. If I hadn't done that, Shirley's head would have been ripped off. Shirley was eight years old and still remembers the incident.

Carbine never caused any more problems.

Fair Treatment

Tommy Rice's greatest innovation was using the car battery to run the radio. He would get me to pull the battery out of the car, bring it into the kitchen, put the clips on it and get it working.

It was a big radio, about half as big as a bookcase. We had to be very delicate with the tuning because if we didn't set the dial precisely on the station we got all these crackles and 'wows' and funny old sounds. News was the main reason it was switched on. We didn't spend any time listening to music or radio plays. We listened to the ABC-News at 7.00 each night, then as soon as it finished – *off.*

However, I was an avid reader.

I read everything I could get my hands on. If it were written, I'd read it. Labels on tins, anything.

I still read a lot.

Tommy Rice had a piano and I was allowed to play it only once and I made a mess of it, so that was the end of my piano playing.

Out of school hours I was pretty busy working all the time, but I was the happiest I had ever been. I drove tractors, trucks and cars from the age of eight. I was driving horses too, though I didn't have big teams in those days.

I like fixing things, making things work and improving things.

I was even a member of bicycle gang, and had a lot of fun with my friends. We played football and rode our bikes – even as far as Parkes, some 50k away. Our gang was known as the Young Kingfishers because one of our member's surnames was Young, another was King, and mine was Fisher.

There were six of us in this gang – Hec and Cobby Young, Bert Leary, the Frecklington boy (I forget his name), Bert King and me – Frank Fisher.

There were a few advantages – not the least was the element of stability. Furthermore, Tommy treated me fairly – even generously at times. For example, he made me a present of his dead son's stamp collection, although I didn't have a clue what stamps were and why anyone would want to collect them.

More significantly, on one occasion he paid me 3d for having worked all day cutting back the cotton bush weed. This was the first money I ever earned.

He also passed on many principles that have guided me throughout my life, and many of those principles were straight out practical things to do with farming procedures.

But most of all, being on that farm gave me my first opportunity to discover that I have a talent for fixing mechanical equipment. It happened when I noticed a dead T-Model Ford in Tommy's back shed.

I was only 10 years old at the time, but after looking it over, I reckoned I knew how I could get it running again, which I did. Tommy was enormously impressed, whereas for my part, instead of weeding the vegetable garden, which I hated doing - I got enormous satisfaction from looking around and fixing other broken-down farm machinery, like reapers, binders and cream separators.

And then I built a cubby house for Shirley.

She probably thinks I was a fantastic brother – saving her life, building her a status symbol to impress all her playmates – but I didn't do it for her.

I did it for me.

CHAPTER 3

47 Odd Jobs

In the years before 1938, during and straight after the war, I had 47 odd jobs. I worked hard for very little money.

One of my jobs was watering the plants. I had to cart water in four gallon kerosene tins with wire handles which I didn't like doing because the handles used to cut into my hands. I also didn't like getting down on my hands and knees and weeding the gardens. To this day, I still don't like weeding. Although I didn't realise it at the time, my future interests would be engines and mechanical implements.

I'm not crying about it. I never thought the world owed me a living. I had no ambitions at all. I didn't worry about yesterday terribly much, and I didn't worry about tomorrow. The day you woke up was the day you worried about. I lived one day at a time; we all did.

The Bottom Bunch

In those days Australia had three classes of people – the upper class, the middle class and the lower class – and I was in the bottom bunch. I don't ever remember being concerned about it. We didn't object. We just accepted it.

I was reminded of this attitude when I went to England many years later and I noticed that the working class people are quite content to call themselves working class and be recognised as such. They didn't aspire to be middle or upper class. In 1938, Australia was very much the same way; we had three classes, and we were the working class.

Furthermore, the working class had levels and a divorced woman with two kids was fairly low down on the totem pole. She wasn't looked up to; she was very much a servant. And Shirley and I were down there too. But we never really thought about it at the time. It was simply our 'place' to do certain things, and we didn't complain.

We didn't have much before the war. Like everybody else, I nearly always wore hand-me-downs. We wore corduroys on very special occasions; other times we wore bib and brace overalls, or shirt and trousers. And we all wore blue singlets, probably the famous Chesty Bond singlets. (The only time I remember wearing special clothing for any purpose was later in life, when I was shearing. Shearers have to wear special soft shoes, heavy trousers with dark blue heavy singlets or sleeveless flannel shirts for protection against burrs.)

Although tractors had been introduced into the wheat-growing areas in the 1920s and 1930s, it was not until the 1940s that they became common on most farms. The same with radio – just about everyone in the world got one before country Australians. If somebody had a radio, we'd sit there and enjoy it.

Parkes

Some time in 1937 Mum left Tommy Rice's Waterford Station and she, Shirley and I, moved to Parkes. Mum got jobs wherever she could. She worked as a farmer's cook, she also worked as a shearers' cook for a while, and later on Mum got a job as a second chef – or pastry cook – in Tattersalls, the best hotel in town. On top of that, she also did some dressmaking, so Mum worked pretty hard. A single woman with two kids did whatever she had to do to survive.

Like Peak Hill, Parkes was originally a gold town. A 'canvas town' (named Currajong) was created in 1862 after the discovery of gold, and nine years later the area was one of the best gold-producing areas in the country. Shortly after, it was renamed from Currajong to Bushman's Lead and in 1873 officially named in honour of Sir Henry Parkes, who was the Governor of New South Wales. Clarinda Street – the main street – is named after his wife.

In the centre of town is a busy roundabout from which the four main roads lead north, south, east and west: north to Dubbo via Peak Hill; south along the Newell Highway to Forbes; east to Sydney via Orange, Bathurst and Lithgow; and west to Condoblin and ulti-mately to Broken Hill. On the corner of this intersection stood Tattersalls, the biggest hotel in western New South Wales. In its place today stands the Coachman Hotel/Motel.

Shirley and I sometimes stayed in boarding houses. (We were boarded out when Mum was sick.) Some boarding house owners, like Mrs Casey, were grouchy, others, like Mrs Davis, were nice. One day Mrs Davis baked a birthday cake for Shirley. Shirley had never had birthday cake before.

However, even though I was living with Mum and Shirley on and off, by this time I was forced to be quite independent, because I never knew where I was going to be housed from one week to the next. I might be out on my neck and sent to some other relative somewhere else. I'd sometimes wear out my welcome if I weren't making a decent contribution to the running of the household. I was probably lacka-daisical about 'paying board' with Mum or anybody. All my life, I had to pay my way as part of the contribution to the family I was living with at the time. I found it frustrating always having to 'earn my keep' because I was so young.

School

At first, Shirley and I attended Parkes Primary, the local public school. We had no religious pressure on us in those days because we were not in a position to go to church very often. We were never required to attend church unless we were staying in a home of religious people. More often than not, we lived on farms that were way out of town and we didn't have any transport.

I was in fifth class when Monsignor Moran, the head of the diocese and our local parish priest from St Jarlath's church, got to hear that Mum was a Catholic. He suggested to her that Shirley and I should attend the convent school. And so when I was 11, I spent 1937 being taught by the Sisters of Mercy (in fact, they had no mercy); there was one building for the boys and another for the girls. The nuns – especially redheaded Sister Vincent – were very fierce. We nicknamed her 'Vinegar' for obvious reasons. I got more canes that year on my hands for not knowing about religion than I got for anything else. It didn't seem to be much of a way of teaching about God.

None of this made any impression on me. I can't stand church rituals, I can't understand Latin and I have never had any time for organised religion. I have since found a record of my Confirmation, but no record of my Baptism.

Religion and I are still at loggerheads, and if the nuns made any impact on my thinking it is only that I believe the world would be a better place if we all lived by the principles of the 10 Commandments. It is the moral basis of human success.

And I would like to be remembered as a success, not just in business but also as a husband and a father – in short, I would like to be remembered as a good human being.

Speedwell Pushbike

My life in Parkes was pleasant enough, because I got to 'play' (which was unusual in my childhood) and also because I had a friend, Frank Stack, who told me later that he had a crush on my sister!

One of the best things about this period of my life was that I had my own transport, a Speedwell pushbike that I'd rebuilt. Someone found the frame and wheels in the weir on Waterford Station, and Tommy Rice let Mum take it with her when we moved out. Together, Mum and I managed to get seats, tyres, pedals and all the missing components, and I rebuilt the bike – with a few 'extras' too, like a shanghai on the handlebars, and a tow-bar to pull a little two-wheel trailer.

Stacky and I rode our bikes everywhere. We fired quandong nuts at people from our handlebar shanghais and we attacked rival gangs, like Dirty Dora's. I rode that bike all over the Central West, from Dubbo to Wellington, Forbes, Peak Hill, Narromine and Dubbo. Apart from the sheer enjoyment of being in a 'pushbike gang' (of two people), it also transported me to better paying jobs.

At the close of the 1938 school year, I passed my Qualifying Certificate (QC) which enabled me to graduate to high school.

Mum Remarries

On 4 May 1938 Mum married David Whitcombe in Parkes. Looking back, he wasn't all that bad.

He and I didn't hit it off terribly well. Dave was not very successful at keeping jobs, though I suppose in his own way he tried to do his best for us.

When Mum met him he was working as a cellar man at a local hotel. Dave had never been married before and so it was a big thing for him to marry Mum, because he not only acquired a wife, but also Shirley and me.

David hailed from the town in Wales that has the longest place name in the world. Before migrating in the early-1930s, he had worked in the Welsh mines and I have no idea why he came to Australia.

I don't think Mum was the greatest person to live with. She was always tense. She created turmoil in everything she did. She would enter a room and the air would turn to ice as she walked through grumbling. Mum behaved as if she thought there was nothing better to do than to cause an argument and she hurt my feelings many times with some of the harsh things she said to me.

David had a pretty unhappy life with Mum. And I probably didn't contribute to his happiness either. I suppose there could have been a bit of jealousy there, but I admit that I wasn't very cooperative. He got on better with Shirley than he did with me. I would sometimes hear them singing together while they were washing the dishes when he was home on leave. They both had nice singing voices.

Employment

Many of Australia's labour laws were being established during 1936-37. These were the times when the 40-hour week was under debate (at that time it was the 48-hour week), paid annual leave (from one week to two weeks) was included in the worker's award for the first time, and prosperity loadings were added to the basic wage.

Employment was more available in Sydney. Out west the farms were fighting diseases in lambs and surviving the droughts, and also a 'falling disease' in cattle. We still had job shortages and many people – like Dave – were on the dole. Due to unemployment, we'd been living tough. We'd been on and off the dole for years.

I only spent two days in 1st year (year 7) high school and then I had to leave to get a job because Dave lost his as a cellar man and Mum was out of work because she was constantly sick, so I was the only breadwinner. Although I didn't know it at the time, Mum probably had morning sickness.

I never had a problem finding work. I left high school when I was offered the opportunity to convert my Saturday job at Burch's General Store into full-time employment. That's how I came to work for old penny-pinching Jack Burch as a 'useful' for 15/- per week – and it was a six-day week.

Burch's Store sold everything – vegetables, clothes, household goods, food and groceries. He not only stocked as many different items as possible, but he'd also get as many hours as he could out of me, without paying me a penny more for overtime. I would start at 8.00am and work all day, sometimes until midnight.

My first task each day was to polish the big wooden floor with a heavy industrial floor polisher. It had a mind of its own and would take off in any direction whenever the balance was altered. So I secured a half-full Vacola bottle of water across the top of the motor which counterbalanced the machine's sudden lurches. And that worked quite well. I could do the job in half the time because I wasn't fighting the machine.

Jack got me to do the heavy carrying work. When the farmers came to town to purchase their supplies, they needed the goods carried to their trucks. So I would pick up their purchases and carry it to their truck, and I would have to do that back and forth until they were loaded up. There was no upside – the work was tough and the wage was low.

In the end, Jack Burch's penny-pinching ways came to the attention of the Department of Labour. When they visited his premises to audit his books, they found that he was underpaying me and they made him back pay me an extra £6.

This he did; then he sacked me on the spot.

I felt this was most unjust, though the £6 was a princely sum.

I gave all the money to Mum. She bought some clothes for herself, Shirley and Dave, and gave me back £2, which I spent on the first suit I ever owned – with shoes to match.

I was quite pleased with myself on that day.

Jobs, Jobs, Jobs

After leaving Burch's, I found work on various farms around Parkes. Some were good paying jobs, but others were not. The pushbike enabled me to chase better-paying jobs five to 10 miles out of town. But they didn't always pay in cash; sometimes they paid in kind. This didn't help with rent, electricity or clothes, although it kept us fed. Sometimes I would be planting potatoes, other times I'd be planting peas, onions, celery, shallots or other vegetables – or harvesting them. I did that for a time.

Next, I worked as a 'useful' at a local maternity hospital. My job was to milk the hospital's cows, keep the yards and public areas clean, weed the gardens, clean up the operating theatres and help the nurses with their extra-heavy patients. I worked a six-day week for 15/6 per week, and they gave me two meals a day – except Sundays – when I didn't go to work.

My last job in Parkes was at Finn's Pharmacy where I cleaned the store, washed bottles and delivered medicine around town on my pushbike.

The first of my two half sisters, Janet Anne Whitcombe, was born in April 1939. Shirley was sent away to Uncle Rowley's station at Tomingley for the occasion, while I just kept on working.

However, in 1939 there was a new mood in the air. Dave got a job as an ambulance assistant and Mum got the job as second cook at Tatts. I quit my job at Finn's Pharmacy so that I could look after Janet while they were at work.

Mum and Dave worked real long hours, and they sometimes only came home to sleep, change their clothes and get back to work. I was in charge of the household chores and babysitting Janet. Some afternoons I would take her to see Mum at Tatts and we would make ice cream together. In those days ice cream was made with a hand mixer. I don't know why they didn't put a little electric motor on them.

It felt as if the employment level was rising. Yes, there was a new mood in the air all right.

By September, Australia would be at war.

CHAPTER 4

War!

I was 13 when somebody told me about the war but – like everybody else in the community – I was less bothered about it than by our immediate problems. I probably should be saying how gloomy things were in the lead-up to such a terrible war, but they weren't gloomy at all. Things were looking up! In the months immediately preceding the war Dave got himself a job. That was *good*. Other people we knew also went and got jobs – frequently in Army outposts or making supplies for the Navy or the Air Force.

The most important thing to us out west was that most people had jobs and money in their pockets. Before that we didn't have much cash money because we'd been living on the dole and in those days the dole was only coupons. We had a coupon for a pound of butter and a coupon for a coat, etc ... but suddenly everybody had money in their pockets and it wasn't just us! We got five bob a day, which we thought was quite a lot of money.

I didn't really understand much about the significance of the preamble and beginning of World War II. However, I thought that whatever we were doing as a nation was the right thing to do.

'King and Country'

I didn't hate Hitler then; I hate him now. I dislike him more with hindsight than I did at the time because in those days we didn't know what he was doing. We didn't have world news. Even though I always loved reading, by the time we got a newspaper it was days old so we didn't know what was happening. We couldn't afford a daily newspaper; it cost tuppence! Anyway, in country Australia, we were too busy living.

In Peak Hill the sheep were more important. I never expected Peak Hill to be invaded. And I think most people thought the same way.

I don't think most of us wanted to enlist to fight for 'King and Country' or out of any great feelings of patriotism (as they did in the First World War). Most of the people that I knew joined the Armed Forces because 'it was the thing to do'.

Furthermore, there was nothing especially noteworthy about the war at the start. For the first year or so they called it the 'false war' because it was pretty quiet and very little was happening.

I didn't have any feelings about the Australian Prime Ministers of the time – Robert Menzies and John Curtin. And I wasn't interested in the English Prime Minister, Winston Churchill. I knew more about Benito Mussolini than I did about Adolf Hitler because Mussolini had invaded Ethiopia, which in those days was called Abyssinia. We had some sort of a rhyme about Abyssinia, which is why we knew about it. We also knew that the Italians invented the first of the miniature submarines and scuba-gear. I don't know why we knew about that, when we knew little else about the war.

I remember the day my stepfather Dave enlisted. Dave had been in and out of work, so when the war started in September 1939, he was one of the first through the door. He joined the Army immediately and applied for a job as a batman to which he was accepted.

Dave spent his whole time in the Army as a batman to senior officers, which must have been a lot easier than fighting with the Rats of Tobruk or on the Kokoda Trail. Towards the latter part of the Pacific war, Dave spent quite some time in Singapore and Malaya. In fact, he was on board the last hospital boat out of Singapore before the city fell.

Sydney

Mum, Shirley, Janet and myself accompanied Dave to Sydney for his induction. We stayed with people in Bridge Road, Glebe. We went back to Parkes for a little while, then we came back to Sydney and we stayed in Seymour Street, Croydon Park, in a house owned by people called the Moffits, also from Parkes. The husband had also joined the Army. Dave met him at their induction at Holsworthy, which is how we got the Croydon Park address where we stayed a while. Mum, Shirley and I lived in Moffits' house, while Dave was doing his basic training.

Meanwhile, life at home was okay. There was a constant stream of visitors. For a while cousin Lola stayed with us, sharing the back verandah with Shirley. I think she was sent to our place because her Mum (my Mum's sister Lila) and her husband were having marital problems.

Ice Cream Boy

In Sydney I worked all sorts of jobs, including assembling padlocks, fixing tyres, cleaning and polishing motor bikes, general dogsbody at a garage, and selling lollies and ice creams at the Palace and other movie theatres.

The ice creams were in small paper cups with little wooden spoons and we would carry about 100 on a flat wooden tray with sides about the same height as the cups. We carried our money in a small flat tin. Sometimes if more than one interval fell at the same time, we would increase the number of ice creams to three trays. This meant we were sometimes carrying more than 300 ice creams.

The Palace Theatre was originally built as an opera house but it was converted to a picture house. It had front and back stalls on the ground level, a mezzanine level – a circle level, a bleachers level, and a very cheap level referred to as the gods.

On one occasion when I was carrying three trays I had to go up to the gods to sell. I was walking up the aisle after the lights were put out and a man jumped up, collided with me and knocked the three trays forward. So, 280 ice creams and all my takings cascaded over the patrons in the four levels below.

It took more than 30 minutes to clean up that mess. While most people were understanding, and returned some of my money, it cost me more than two weeks' wages for dry cleaning and lost sales.

Despite the mishap, I later got promoted from Ice Cream Boy to Box Boy where I manned a booth, selling three grades of confectionery boxes: one with sticky-jaw toffee, one half-toffee and chocolate and the real fancy boxes of chocolates which the Americans bought. They were the only people who could afford them.

It was great having the customers coming to me instead of me having to carry the heavy lolly or ice cream trays around the theatre where I might risk dropping them on the patrons – again.

I did three sessions per day: 10/11.00am, 1/2.00pm and 7/8.00pm. During the three hours I had to spare before the 8.00pm session I would go to Lindrum's in Pitt Street, which was our nearest snooker parlour, and I was taught the game by world champion, Horace Lindrum himself!

However, I was never very good at the game because I was shortsighted.

Smoking

In 1941 I worked at the Eveleigh Railway Workshops for 14 months as a 'shop boy' (ie 'workshop boy') heating rivets. The workshop also made gun shells. Thousands of castings lay around the ground before they were to be machined down into shells. We had to walk across them to reach the toilet and that was a problem. One day I tripped on one and sprained my ankle.

We lived in Bridge Road, Glebe, at that time. Mum must have got some extra money from somewhere because she bought me a banjo-mandolin. She felt sorry for me when she saw how unhappy I was about staying home with the sore ankle.

The banjo-mandolin is a melodic instrument and I'd strum its double strings. Somebody gave me a book on how to play it and I learned all the popular songs of the day, like *Way Down Upon The Swanee River*, *Lili Marlene*, and *White Cliffs Of Dover*. I've always had a bit of an ear for music. We'd listen to Vera Lynn and other singers on the radio and we'd pick up the tunes that way.

Being in Sydney gave me a chance to get around more than I used to when I was living out west. I had money in my pocket, I had a job, I owned a suit, and I often went to dances. That's how I came across Jack Huxley, who taught me how to smoke when I was 15. We met at a dance at the Petersham Town Hall and he became a good friend.

Having learned from Jack, I went off and bought my own packet of cigarettes before catching the bus home. The brand was known as 'Three Threes'.

In those days, the top deck of the bus was the smokers' deck. And I headed upstairs, sat in the front of the bus, took out my first cigarette, lit it and thought I was big time.

After a few puffs I started turning green, and then I had a sudden urge to vomit – and it was a long way from the front of the top of the bus to the back door where – to everybody's disgust and my extreme embarrassment – I did the big spit.

I should have given up then because it was the ideal time to do so, but I persevered because 'it was the thing to do'. If you didn't smoke you were a strange person, and I was strange enough already because I did not (or could not) drink beer or other alcoholic drinks.

Unsuccessful Con Man

And then my father walked back into my life.

It happened this way.

My father owned a company called Gilbert E Fisher Pty Ltd Advertising while he lived at 99 Victoria Street, Potts Point, where he told me he was after he separated from my mother.

Before the Second World War, all lending libraries were privately owned, and you could rent a book for 3d a week. Every shopping strip had its own lending library.

My father printed library bookmarks for a living. (I still have one; it's the last one I have left.) He would call on shops close to the library and convince them to advertise on his bookmarks for 25 shillings for 12 months. He would then print them up and the librarian would insert one in every book that was lent out. It appeared to be good advertising.

Mum's sister, my Auntie Lila, had a dress shop at Bondi Beach, and one day my father decided he would go and sell some advertising to that area. My father had called on all the shops around Bondi Beach and had offered them an 'advertising opportunity' on his bookmarks. He walked into her shop and they both realised that they knew each other. She immediately recognised him as her sister's ex-husband. He wasn't difficult to identify. He wore a suit and tie, smoked cigarettes through a cigarette holder, and his manner was always charming – a real con man.

If he can be believed, my father claimed to my auntie that after he and Mum had broken up he had pursued every avenue in order to try to find Shirley and me. He was certainly keen to see us again.

But Lila feared getting Mum's hackles up, so a secret meeting was arranged. I was contacted at work, at the Eveleigh Railway Workshops, and I met my father at Auntie Lila's shop at Bondi Beach, then we crossed the road and went swimming.

That's about all really.

Nevertheless, Shirley and I never lost contact with him again.

Dave and EMI

Meanwhile, Dave was de-mobbed from the Army in 1941. He was deemed medically unfit after suffering severe tinea brought on in the tropics of Singapore and Malaya. He brought with him all sorts of stories about military activity in foreign lands, which seemed quite alien to what we were experiencing back home.

I was in Sydney when the Japanese miniature submarines came into Sydney Harbour. First, we heard that they had raided Darwin in February 1942. Then, in May, three Japanese midget submarines attacked Sydney Harbour and fired a torpedo that sank an ex-ferry, the HMAS Kuttabul, and killed 19 young ensigns who were sleeping on board. I don't remember any other military activity on our shores.

Dave got a job as a storeman at EMI/HMV on the Parramatta Road, Homebush, across the road from the Arnott's Biscuit factory. He stayed in that job for the rest of his working life.

Dave also got me a job at EMI. He told someone there that I was interested in radio, so they gave me a job as a 'press operator' in the pressing department of the radio section, which isn't exactly what I meant! I was bored out of my brain within two minutes of starting.

In those days, valve radios were built around a square metal chassis with lots of holes in it. It was my job to run the wire through the various holes. Later I was promoted to a radio chassis maker, which was equally as boring. I had to pick up a piece of metal, put it on the press, close the front, press the button and it'd all come together into a square box. Then I'd pick it up, put it on the heap, find the next one and do the same thing all over again: pick up a piece of metal, put it on the press, close the front ... over and over again, all day, every day.

I should have quit immediately. I was still only 15 and I had that right. However, I stayed and turned 16 and was then subject to the Manpower Act.

During the war years of 1939-1945, the Government introduced the Manpower Act that prohibited anybody over 16 from quitting their job. It was a very harsh Act. This legislation gave the Government the right to place anybody in whatever job the Government

War!

wanted them to do. There was no choice. It was a dictatorship. I couldn't just walk out. It was a criminal offence to quit.

I was going off my brain! I was used to Peak Hill. Instead, here I was in Sydney, working at the most boring job in the world – and I wasn't even allowed to quit. Mr Brown was my boss. And Mr Brown and I didn't hit it off very well. Even though I gave him my notice, there was no way in the world he was going to let me go because he was desperate for staff. This left me with only one option: I had to find a way of making him want me to leave, so I became an 'unruly person'. I kept on breaking dies and making messes; I distinctly set out to get myself sacked.

One day I had to put this thing – which was about the size of a matchbox – into a press. Then the press came down and put holes in it.

It was very delicate and very valuable, being made of gold. Every part had to be guaranteed and picked up afterwards.

This repetitive work can send you off your head, and I put two on the press at the same time.

I then pulled the press that came down with a Ssssssssh.

It broke the die and scattered the two gold casings.

He sacked me.

From that time onwards I was never very popular with Dave.

Riverview Station, Peak Hill

When I was 16, I went back to Peak Hill. Among all its other attributes, Peak Hill was a popular place for resting racehorses. Quite

57

a few stations there used to look after horses when their owners wanted to put them out to pasture for a month or two.

I got a job for 15 shillings per week at Riverview Station, which was owned by Billy Swain. Billy Swain always had many horses at his station and one of my jobs was to look after them. When he realised that I wasn't interested in horses at all, he taught me to be a butcher. He taught me how to kill and butcher sheep, pigs and steer. They were each different – you kill a sheep by cutting its throat, you kill a pig by stabbing it in the throat and you shoot steer.

After butchering a pig, the meat went into a drip safe (some people call them Coolgardie Safes) where the water ran down the outside, dripped through hessian and the cool air kept everything from going off. Chops, you'd eat as fresh meat, but we'd put the other parts into brine, leave it for several days and salt it. Ham and bacon were put into the smokehouse to cure.

I didn't particularly like being a butcher, I didn't like the work, and so I quit and hired myself out for any kind of work around the district. However, even though the years have passed and Billy Swain isn't around anymore, I have good memories of him. He was a patient man who taught me many things and it wasn't his fault that my skills lay elsewhere. He did his best.

All my life I would read anything that came to hand. During the war years reading matter was simply what you had. I read many biographies and I particularly enjoyed books about mechanics, science and science fiction. Up-to-date newspapers were scarce, so I read whatever was in people's houses. As a result I read the Bible from cover to cover several times – the first time was at Billy Swain's Riverview Station, where it was the only book in the house.

I am pleased to say that the Riverview Station is still standing today.

Billy's son Stephen Swain and his family live there and run the station.

Odd-Jobbing Around Peak Hill

There weren't many men available in those days to work on farms. I would travel around the countryside of Peak Hill, Parkes and other western districts. I worked on a person's farm as a contractor for an hour, two hours, five hours or a week – doing whatever they might want me to do. And then I'd go off and do another one. I had about 40 contracts.

Whatever job needed doing is what I did for a living during the war years. I drove a tractor at night with a hurricane lamp hanging off the crank handle. I did crutching (cleaning the backsides of the sheep so they wouldn't get flyblown). I put in rows of fences. I dug dams. I helped with wheat stripping. I ploughed fields.

In those days they used to put the hay into stooks. I did stooking. I would bind them around the middle and then 'stook' them up (stand them up).

I did rabbiting.

I cut chaff.

I made hay.

I drove truckloads of wheat to the silos in the 1936 Bedford truck.

All different types of families would put me up. One example was Bill and Bert Job. They were religious and insisted on my attending church as part of conditions of employment at their Kidina Station.

Bill and Bert were brothers and First World War veterans. They were given 2000 acres of land each after the war. It was called a 'soldier's settlement' in that part of the world.

Bill and Bert married twin girls. These twin girls were very large ladies, and both Bill and Bert had very large families. Bill had 18 and Bert had 16 children, or vice versa. Even though they had lots of children, they needed people to work for them – because they had their children in a hurry, and they were all babies or small children.

Apart from attending church on Sundays, my main job with them was ploughing, driving 16-horse and 18-horse teams. I also used to have 10-horse teams for pulling loaded wagon.

I left Bill and Bert to go back to working for Uncle Rowley, Big Auntie Violetta's brother. He was one of the relatives who put me up and fed me when I was a child, so I owed him a debt of gratitude. There was a shortage of manpower because of the war, so I helped him for a time. It was pleasing for him to see how I had grown, and satisfying for me to be of use.

I Wanted To Enlist

By this time, everybody I knew was in the Armed Services. You felt wrong if you had not enlisted. Australia was fighting Japanese troops in Malaya and New Guinea. The sloop HMAS Yarra had been sunk, as had HMAS Canberra. And the War Cabinet ordered 105 Commonwealth Aircraft Corporation (CAC) Boomerang fighters to be built. It was all exciting stuff!

I wanted to join the Army, Navy or Air Force, not because I hated Hitler or the Japanese, but because it was exciting – it was the fashionable thing to do.

You could join the Navy at 16 so I tried to join in 1942 as a cadet and I was surprised to get knocked back as 'physically unfit'.

The problem – I found out later – dated back to an accident that had happened when I lived in a hall on the corner of Talbregar and Macquarie Streets, Dubbo. One day I came running around the corner in the dark. Somebody had closed the gate that was not usually closed, and I ran into it.

I broke my nose – *choonk!*

I didn't go to hospital, we couldn't afford that; I went to bed until it stopped bleeding, that's all. It hurt like hell.

In those days you were rejected from the Services if you couldn't breathe through your nose. Apparently, when it healed, the nasal passage closed up, so I mostly breathed through my mouth instead. I didn't even know about it.

Yet, I was an otherwise strong boy. I could carry three bags of wheat on my shoulders (I have a photo to prove it). I could drive a tractor and a car. I was quite capable of doing most things – so I was shocked when they told me that I was supposedly physically unfit for the Navy, and they said the same thing again when I applied later.

So I shrugged my shoulders and went back to the farms.

Flying High

After being knocked back for the Navy, I joined the Air Training Corps (ATC) and I was accepted. The ATC was a group of young men aged 16 to 18 who were Air Cadets with full Air Force uniforms, who trained as airmen and who were expected to join the regular Air Force after reaching 18 without having to go through the normal Air

Force training. We had a fighter squadron at Parkes, and we became the Peak Hill Flight of the Air Training Corps.

During the war years the Government had 'war bond drives'. One of their promotional efforts was to exhibit a Hurricane fighter aircraft under the command of a pilot who had been through the Battle of Britain. They took the aircraft around the various towns so people would see it, get enthusiastic and buy war bonds. That single-seater Hurricane fighter with just enough room to cram in two people, including the pilot, came to Parkes.

The pilot addressed the crowd, after which he said to us in the Air Training Corps, "Step forward anybody who would like to have a flight."

Naturally, we all stepped forward. So he picked up me and another kid and we went on a 10-minute flight in his Hurricane.

The pilot asked me, "Where do you live?"

I said, "Peak Hill."

He said, "Okay" and blasted the engine, and off we went for our 10-minute flight. (It's 30 miles from Parkes to Peak Hill, as the crow flies.) It was quite unbelievable; it was so hot and noisy. Oh God almighty! Can you imagine sitting up there with this engine with no mufflers, just roaring its head off? This thing went up like a rocket and came down like one. It was the most exciting moment of my life.

We were the last people to fly in that plane. When they checked it before its next flight, the aircraft was declared to be unserviceable and was grounded. Later it was partly dismantled and transported to different parts of the countryside by truck.

Peas!

During the war years the Governor of New South Wales, Baron Lord Wakehurst, visited Peak Hill and part of his visit included a public reception at a community hall where he was invited for a meal. I was in the Air Training Corps at that time, and therefore in his guard of honour, standing with all the other cadets in a position where I could get a good look at Baron Wakehurst's eating habits. I was curious about this after all the fuss Uncle Edgar made when we were little, about how to eat peas. I knew peas were on the menu, so I was very interested to watch what the Baron would do.

Finally, the official party sat down, while I and the other cadets lined the walls. Nobody could start eating until Baron Wakehurst started his meal. It was particularly interesting to watch him eat his first mouthful of peas.

He held his fork upside down! Yet Uncle Edgar had taught us to lift the peas on the back of the fork, not scoop them up like a spoon.

I couldn't believe my eyes when the Governor of New South Wales did it wrong!

CHAPTER 5

Travellin' Around New South Wales

When I was 17 I re-applied to join the Air Force, and once again I came to Sydney for my physical examination and was again rejected as physically unfit. So I went back west and spent the next few months working for Ignatius Aloysius Fitzpatrick McNamara – 'Nacy' for short – a shearing contractor from Tomingley. He had a line of contracts which started in August at the back of Bourke from where we spent the next five months winding our way home through those western NSW towns until we finished up in Orange in time for Christmas.

I started out as a shed hand 'picking up', which means I would pick up the shorn fleece, throw and spread it on the table for classing. I also trimmed all the rubbish off, like dags that I would pull off and throw in the dag bag, and I also did shed hand work – bringing the sheep in and stuff like that.

Shearing

A 'shed' refers to the group of shearers as well as to the actual building itself. The buildings were used only a few times a year, so they were an expensive piece of real estate for the short time that they were used.

A shearing shed generally has a minimum of 'four stands' with four handpieces hanging from a shaft above each shearer. Most shearing sheds have eight stands. Sometimes there would be 12, sometimes 24, but they were always in multiples of two. The shed would also have to employ at least one pick-upper, one wool classer, one presser and someone to herd the sheep. The shearers got paid according to the number of sheep they had shorn, while the others were paid a standard wage.

An eight-stand shed would take about 36 feet of space and the holding pens that were opposite the shearers were about the same size. Another essential part of every shed is the classing table, which is approximately eight feet wide by about 12-14 feet long, made of single two-inch-wide wooden slats about an inch apart. Behind the shearers is the shoot where they send the sheep down to be counted and kept in their pen after they've been shorn.

We mostly sheared merino, although we did have a lot of 'come-backs' which are Merinos which have been bred into Border Leicesters (or something else) then bred back to Merino. This makes their skin smoother and easier to shear, but their wool is not as fine as pure merino.

After working as a pick-upper for a few months, I started shearing. I averaged about 160 sheep a day, which was a modest number. Other shearers were doing a lot better than that, some averaging up to 50 more sheep per day than me.

One shearer was nicknamed 'Mr State Express' after a popular brand of cigarettes of that time. One of their product lines was 'Three Threes', the brand I smoked on my disastrous first smoking experience. We called this fella 'State Express' because he once sheared 333 sheep in one eight-hour day!

The shearer's handpiece drops down from a shaft, which is driven from outside the shed by an engine of some sort, in most cases a simple steam engine. To operate each shed the station needs to have a steam, petrol or diesel engine in working condition to drive the plant. Nacy didn't like fiddling around with the engines, but I was adaptable, so Nacy put me in charge of all machinery instead of shearing sheep.

The shearers' wages were averaged out and I would receive their average as my pay. And that was pretty good because I was good at getting those old machines to work. Some of them were only used twice a year, so there were many poor old decrepit things lying around doing nothing for months. Some of the old oil engines were the hardest to keep going because they run on heavy oil. Some of the steam engines were pretty rough-and-ready too because of rust, but I had to keep them going while the shed was working, because the equipment couldn't stop.

I would arrive at the various stations at least one day before the shearers would turn up, and it was my responsibility to make sure all the machinery worked before the rest of the shed arrived. Shearers don't get paid to stand around, they only get paid by the number of sheep they shear.

In order to travel ahead of the team, I bought my first car, a clapped out 1934 A Model Ford utility, which was never registered. It didn't have to be. I just drove it around the countryside and never into town. That A-Model Ford gave me even more experience in keeping an engine running!

A sheep is shorn this way: the wool around the head is shorn first, then the legs, then the belly – which are small operations – and then the shearer does what is known as the 'long blow'. Starting from the head, the cut is towards the tail in one long blow. If possible the shearer tries not to stop each stroke until the sheep is fully shorn. It is the way a super-shearer earns the best money. Some shearers can shear a cleanskin sheep in three to four minutes. The machinery worked pretty hard to keep eight shearing stands going, particularly when eight shearers were all working at the same time doing the long blow.

The 'gun' of the shed is the 'top gun'. Everybody is aware of what he's doing. Each time a shearer goes into the pen to pick a sheep he always grabs the best one, and there's a fair bit of competition between the shearers. If one sheep has got more wrinkles than a baby and is as old as Methuselah and the other is a young, smooth-skinned Merino ewe – which one are they going to grab? The old wrinkly? Of course not. They are also competing with the other shearers for the best and easiest one to shear, so each shearer has got to keep an eye on what's going on all the time.

Wool classers have also got to reach a high standard of accuracy. No two sheep have the same grade wool, so they are graded individually. A classer has got to be accurate and to instantly identify the difference between an A grade or AA grade or any other grade while doing his work as fast as the fleece are being shorn and spread on to the table.

I Don't Eat Chicken

Unfortunately by 1943-44 a national drought which had begun about the same time as the start of the war in 1939 was taking effect very desperately.

In 1940, for instance, the Nepean Dam in Sydney was dry and most western districts had experienced five or six dry seasons in a row.

Some small relief came in 1941 with a few showers, but by 1943-44, 90 per cent of New South Wales was hard hit, especially in the inland sheep-growing districts. In 1945, 10 million sheep died from the drought. Under those circumstances there was no more shearing work so I stuck with Nacy and transferred to working on his farm at Tomingley, doing fencing and other maintenance work, because Nacy too was losing sheep fast.

Then Nacy decided to have one last try at saving his stud sheep during that drought. He asked me to take his last 500 breeders to the Long Paddock, which was the stock route that anyone could use. I set out alone with the herd, as well as food and water, two dogs, two horses and a big wagon loaded with hay and wheat to supplement the expected feed on the stock route. Its back wheels were six feet in diameter and its front wheels were only slightly smaller. By day the dogs rode in the sling at the back, and by night I slung my bed between the axles.

As I travelled, the sheep kept dying because there wasn't even a blade of grass for them to graze on. Up to 10 sheep or more died each day. As they died, I stopped to skin them. Alive, they were worth only threepence; dead, their skins were worth sixpence. After some six or seven weeks the whole herd had died.

I had left with a wagonload of hay and wheat. I returned with a wagonload of skins.

We had no sheep left on the farm and there was no sense in planting wheat, but we did have plenty of seed wheat. The only course was to feed the wheat to chickens, ducks and turkeys. So we grew poultry instead. When they were big enough to eat, I butchered them and we ate fowl for every meal over the next nine months: eggs for breakfast, chicken, ducks or turkeys for lunch, and the same again at dinner.

The smell of hot water on poultry feathers, or even the taste of chicken, brings back the memory of those horrible days.

I never eat poultry.

1927 Harley Davidson

In early 1944 I decided to leave the central west. I'd already got rid of my A-model Ford utility and by this time I owned a 1927 Harley Davidson motor bike. It ran very well on power kerosene, which most tractors used, so despite wartime shortages, I was able to get enough fuel to go down to Sydney to see my mother and family for the first time in more than two years.

I had bought the 1927 Harley Davidson 7.9 motor bike for £8 at Cumnock, which is a little town a few miles east of Peak Hill. The Harley had been owned by a young man who was fighting with the 8th Divvy in the Middle East.

This was not my first motor bike; I had previously paid £2 for a 1925 Douglas flat twin that had a broken bearing in the front cylinder valve, so the bike wasn't running when I bought it. I made a new bearing and rode it all over the place. I then sold the Douglas for what I paid for it and bought the Harley.

The 7.9 Harley Davidson was a very good bike with the added advantage of having two tanks: one for petrol and the other for power kerosene. We could get kerosene quite easily during the war years, and the bike would do 20-30 miles to the gallon.

When I bought it, I set off from Cumnock to Peak Hill with only one gallon of petrol, and it wasn't enough to cover the distance. At nightfall I ran out and pulled over on to the side of the road where I found a farm shed. I slept the night among the sheepskins.

The next morning I called on the farmer whom I persuaded to give me a pint of petrol and some power kerosene. I was able to roll-start the bike on the petrol then switch to the power kerosene and ride home.

A similar incident occurred when I was riding to Sydney. The Harley stopped when one of its valves got stuck. I stopped at Woodford on the Great Western Highway in the Blue Mountains and slept the night in a phone box by the side of the road.

I left the bike with a farmer, caught a train to Sydney and returned the next day to fix the valve. I also managed to purchase a gallon of petrol that I had bought for 10/- on the black market. There was no other way of buying petrol, because it was an unregistered bike.

(Over the next few years I owned a number of bikes – a Matchless, a BSA, a HRD and much later I even bought an LE Velocette.)

White Feathers

I was very aware of the war years, I can remember them very clearly but I wasn't involved in the war in the same way that most other people were.

I should have been in one of the Services!

People gave white feathers to cowards and some people gave them to me, which was upsetting because I didn't like being thought of as a coward. I had tried to join the Navy, Air Force or Army a number of times.

I had been rejected as medically unfit so I carried my rejection note with me at all times. I had to produce it many times to protect myself from trouble.

On the back of my hands I still have scars from the time I was attacked by three Yanks at Burwood Station in Sydney. I was walking up the stairs and they were walking down. I was dressed in civvies, they were in uniform, and they said, "Hey, why ain't you in uniform, man?" or something like that.

"I'm not old enough," I replied, with which they called me a liar and bashed me up. I was badly knocked around. One of them hit me with a knuckleduster that tore the back of my hands to pieces. I still have the scars. The war years were a strange time.

Murwillumbah

And so I stayed with Mum and Dave for six or seven weeks in Sydney, but I was a bit underfoot. When Jack Kelly, my cousin, told me about his home in the northern New South Wales town of Murwillumbah and suggested I go up with there him, it seemed like a good idea.

Cousin Jack had served in the Army in New Guinea and was about to be de-mobbed. Jack was married, and his wife lived up there, and the idea was that I should live with him and his wife and work for him in Murwillumbah. So I sold the old trusty Harley and travelled to Murwillumbah by train.

Soon after our arrival, Jack's marriage broke up, and I had to find other accommodation. Nevertheless, he provided me with employment as promised, because he had bought an ex-Army tip truck and needed an offsider. Jack taught me to drive trucks and I soon got my truck license.

Sometimes things go right for you in life, and this was such an occasion. I soon fitted into the town and I was invited to join a group of young men (average age 17). I did not drink alcohol, and for the first time that didn't matter as more than half the group weren't drinkers either. Over the two years I spent in Murwillumbah I never saw one of them drunk.

There were nine of us, all about the same age. There were three Kevins in our group. Kevin Felton was my best friend and the youngest – he was about a year or so younger than me.

Then there was another Kevin, plus the eldest one, Kevin Went – Wenty. The last time I saw him he was a police sergeant in Redfern, Sydney. Wenty was a great man to have around to keep the peace, as he was 6 foot 2 inches tall and he weighed some 14-15 stone. Although he was a big man, he was a very calm sort of person. If somebody wanted to pick on anybody in our group, Wenty would just walk over, glare at them and say the word, "No"! And seeing this big man, they said "no" too. Wenty was a nice person to be around.

By this time I was living at Mrs Leggs' boarding house. She supplied only one meal per day – breakfast. There were three boys boarding there: Kevin Felton, a fellow called Hughie and me. Breakfast was at 6.00am and anyone who was late missed out and went hungry.

All the others lived in various boarding houses around town too and we met up daily at Mrs Black's Café in South Murwillumbah for lunch and dinner, six days a week, Monday to Saturday. But never on Sundays.

Mrs Black gave us sandwiches for lunch and a baked dinner every evening. We had to be there at a certain time; if we were late we didn't get fed. It was strictly *boom-boom*; there was no wandering in and out to suit ourselves. We had meat and vegetables and sweets (like jelly and custard). We had to sit down in the café and eat our meal properly.

The only other place was the Bluebird Milk Bar, which was more expensive. The girls used to wear uniforms with the word 'bluebird' printed on top of their breasts and we boys used to get a bit titillated that they should have the words 'milk' and 'bar' clearly written across each breast. It was quite something!

As usual, I was working at various jobs – from truck driving for the local Shell Oil depot to working on the banana plantations. Often I worked with Jack Kelly. We used to cart sand and gravel to the local building sites as well as doing other odd jobs that required his truck.

During the war years you needed coupons to buy some food and most clothing. By the time we had paid 25/- for our food and our board we had no money left over. We carried our clothes around in a cloth bag, like a carpetbag. We didn't leave them lying around or they might have got pinched. Kevin Felton and I never seemed to have enough coupons or money to buy clothes, so we shared most of our clothing. (This did not apply, of course, to shoes and socks or ordinary working clothes.) We had a saying 'first-up best dressed' because we only had one suit coat between the two of us. We would buy trousers as close to the colour of that suit coat as possible because trousers wear out, while a suit coat never does. You don't need as many coats as trousers.

In those days a dance was a collar-and-tie affair. We didn't go to a country dance dressed in casual clothes, so we could never attend the same dance together, because we only had one coat between us.

Sick In Murwillumbah

When I worked for the Shell Oil Company in Murwillumbah, all the petrol came in 44-gallon drums (200 litres these days). These drums were trucked from Brisbane and dumped at the Shell depot that was run by a lady whose husband was overseas in the Army.

I would pick up the drums and take them to one-horse towns such as Uki, Mooball, Burringbar and Minjungbal. They didn't have a tanker delivery up there and I would fill up the tanks; some were in the ground, while others were on stilts. If they were on stilts we had to pump the 44 gallons by hand into these higher tanks, whereas if they were in the ground it was much easier because we just ran a hose straight into the underground tanks.

Later in the war, they started importing the big solid oil drums that were made by the Americans and left in the tropics. They were so much stronger than the ones we'd been using. They had big steel rings, so you could get a proper hold. They were very much tougher and galvanised too, so they didn't rust. When an American makes anything they over make it.

("You want a 44-gallon drum?"

"Sure man."

"You can drop it from 100 feet in the sky and it won't dent!")

However, the drums we were handling were made from a very lightweight metal and designed for one use only. After handling two to three times, they would start to deteriorate and get lots of dents and leaks in them.

I got a scratch from one of these American drums that had come from New Guinea. I caught some sort of ringworm on the face. The infection worsened and after a few days a doctor put me in the Murwillumbah Hospital as an outpatient. He took one look at me and said it was ringworm all right, but it was also tinea (they called it ringworm-tinnea). I finished up with it on the whole of my body except the groin area, on the soles of my feet and in my hair. When I was in hospital I used to have two baths a day in Condy's Crystals and water. I wore a bandage from head to foot like a mummy and I was like that on and off for nine weeks before I was finally discharged.

Murwillumbah is not known for its clement weather. It's a very steamy part of the coast, and the steam and perspiration didn't make me feel too good. Apart from the irritation caused by the heat, I was well enough to work. I got on at the hospital pretty well. This ringworm-tinea was not contagious and I wasn't injured, so I was quite mobile. So I ended up becoming a general dogsbody around the hospital. I changed the bedding, helped with the patients who needed lifting out of bed and made myself generally useful. I'd had a bit of experience of hospital work, when I had previously worked at the maternity hospital in Parkes.

I spent two years in Murwillumbah. It was probably the happiest time of my life until I met Pat. I had a lovely bunch of friends up there; they really were special people.

Accepted, At Last

Despite many rejections when I tried to enlist during the early stages of the war, I was actually called up by the Army in 1945 just before the war ended. I was accepted as medically fit and asked to stand by for my call-up and induction.

Technically, I am still on stand-by. I have never been officially notified that I am no longer required. I still have the letter of acceptance that has never been rescinded. It requires me to 'please stand-by and wait for call-up' and I'm still waiting.

The war ended in 1945, when I was 19. Although Germany had surrendered in May, victory was not won in the Pacific Region until 15 August when Japan accepted the terms of the Allied Forces.

A total of 993,000 Australians had served in the various branches of the armed forces; 27,073 were killed in action, 23,477 were wounded and 22,376 had been prisoners of war.

Mostly, men around my age.

CHAPTER 6

Patricia

The war ended with Japan's surrender and my banjo-mandolin
again came into its own when the township of Murwillumbah
joined in the victory celebrations with singing and dancing in the
main streets.

It was an apt climax to my stay in that northern town.

Sydney 1945

Shortly after the end of hostilities I decided to move back to
Sydney to seek work, so I wrote to my father, asking him to lend me
money to get me to Sydney. He forwarded £5 on the condition that
I worked for him and paid him back.

He ran his advertising/printing company, Gilbert E Fisher Advertising Pty Ltd, from shared premises in Blackfriars Street, Chippen-
dale, that – as I said – sold advertising space on bookmarks to retailers.
He gave me the job of setting up the type and working the foot-op-
erated printing press. And I soon twigged that my father was a lazy
man.

He certainly had an air of gentility about him. He peppered his speech with phrases like, "Under the circumstances..." and "Taking all things into consideration..." which made him appear to give serious thought to whatever he was talking about. But he never really said *anything*.

Even worse came the realisation that my father was a bit of a con man because he would contract to supply 1000 bookmarks at a time, but if he printed 300 that was about tops. He never gave anyone full value for money.

There were further complications too, because working for my father and living with Mum and Dave created a new set of tensions. They didn't like him, and they didn't like me having anything to do with him. However, I needed to resolve a few things in my own mind. He was now not just my father, but also my employer.

I nearly cut off my finger while working for him. It happened like this: the printing press (known as a 'platinum press') had a round table over which the ink rollers would pass and then return over the 'platinum' – or the actual ink type. The two plates would come together, and the rollers would roll over the plate that was covered with ink. Then the ink would dry and the only way to clean it was by scraping off the dried ink with a knife. That was one of my jobs.

To do a good job, the knife had to be razor sharp and one day I missed my cut and sliced off part of my finger. I still carry the scar.

So I decided it was time for me to look for another job, which I found with a furniture removalist in Campsie.

Meanwhile, Shirley married James ('Mick') Crummy on 26 September 1945. She was 17. Mick, who was five years older than her, was in the RAAF. He came from a wellknown family in Cunnamulla in south-western Queensland and they set up their first home at Lemon Tree Passage.

Mum Dies

In 1946, my second half sister Gaye Frances, was born. However, Gaye saw less of Mum than any of her other children, because Mum died of a cerebral hemorrhage within two years.

I don't remember how I came to be living back at home, but at the time Mum died, that's where I was. I remember the marriage was going bad, and Dave – ever a tippler – was drinking six schooners of beer every day, which displeased Mum and led to even more arguments.

On one occasion Mum dropped in quite unexpectedly on Shirley and Mick's place at Lemon Tree Passage. Mum reckoned she needed a rest because she was having bad headaches, but Shirley reckoned it was a short separation. On that occasion, Mum stayed with Shirley and Mick for a week.

Shirley's first child – Neil – and Mum's last – Gaye – were approximately the same age. Neil was born in August and Gaye in October, so Shirley and Mum grew closer during the last two years of Mum's life. I was staying with her and Dave at the beginning of September when – while she was doing the washing – Mum felt a sudden agonising pain in her head and was rushed to the Prince Alfred Hospital.

I still wonder whether it related to the time that she fell down the stairs, when I was three years old.

We all visited Mum in hospital. On our last visit I must have sensed something was not right because as Dave, Janet and I were leaving, I turned back to give her an extra kiss on the cheek and to say, "Get well, Mum."

After we left, she died within the hour. I was moved to realise that sometimes the little impulses that impress us all can make us do good things.

Mum is buried in the Presbyterian Section of the Rookwood Cemetery in Sydney. I am puzzled about this, because she was Catholic. However, both my father Gilbert and her second husband Dave were Presbyterian, which is the simplest explanation, I guess.

1946-1947

At this time I was living at a boarding house in Belfields and working for a furniture removalist at Campsie. I was employed as storeman and assistant, which required a lot of loading and unloading of trucks, containers and stacking furniture. It was an opportunity to learn about the efficient use of space and orderly ways of doing things.

Eventually, I became a truck offsider and driver with George Wadell. He was the No 1 driver. George had been a child prodigy on the violin and could still play, but he had become an alcoholic. I think the company put me with George because I didn't drink. We were a great team, handling the special jobs and interstate work.

Sometime in 1947, the company secretary decided to leave and set up his own removalist business. He offered me a partnership that I foolishly accepted, contributing the small truck I had purchased to carry cartons and packing items. I was required to quote on jobs as well as transport small items. His contribution was twofold: to contract the large pantechnicons used to transport the furniture, general management and second – to my detriment – handling the finances.

I had been conned. It wasn't long before we went broke. I was forced to sell my little truck and look for other work, while the other bloke went on to start a new and ultimately successful removalist company.

So at the age of 21 I got my 42nd job as a tyre repairer at Wilky's Tyre Services. For my 43rd job, I became a motorcycle deliveryman, all of which was definitely unfulfilling work, so I started looking out for something better.

'Country Expert'

In 1948 I saw an ad for a position as a 'country expert' at Cooper's Engineering and I applied for it. (Cooper's Engineering was later taken over by the Sunbeam Corporation.) I was familiar with Cooper's Engineering because a lot of engines, shearing plants and dipping plants were made by them and were used throughout the country.

The advertisement intrigued me. I applied and found out my job would be installing home lighting, sheep shearing and dipping plants on country farms and stations all over the State. Some of these plants required someone with a fair amount of farming equipment knowledge and I was the ideal person for that job because of my background experience.

As I got to know the job, I was often frustrated by the farmers, because in order to do proper installation of the home lighting plant and the batteries, they needed to construct a dedicated room to do the job. However, many farmers would take shortcuts and ask me to put the plant in the cheapest and most convenient area, instead of where it should be. Although I was not a licensed electrician, I had to install the wiring to Australian standard specifications (which were known as SAA rules), to which all installations had to conform.

In those days we had red, black and green wires which we had to put through a metal conduit. The conduit had to be earthed, therefore each piece of conduit had to be clean (and paint-free) to allow for an actual earth contact between each piece of conduit. That made a second safety factor. In those early days, we put in 32 or 50 or 110-volt installations.

I spent nearly two years 'experting' around the country. I had a driver's license but the company didn't provide us with a car. I didn't own one, so I travelled by train. My itinerary was all made out by head office. I'd go to a town like West Wyalong – which is between Cowra and Griffith – and the person who ordered a plant would meet me at the station and give me accommodation for the three to five days that it took me to install the plant. After completion of the job he would take me back to the station and I would catch a train to the next place. I would be away for six to seven weeks at a time, then I'd return to Sydney for one week's leave.

Living such a disjointed existence, I did not rely on Sydney for my social life, though I did enjoy my short stays in the city. Most of the time, however, I was living on whatever farm I was assigned to, to install the Cooper's plants.

While in Sydney I would stay at a good friend's home – Keith Smith. Keith lived with his Mum in Campsie. Keith's mother gave me a little room to sleep in and a place to store some personal belongings. (He was best man at my wedding. I still occasionally see Keith and his wife Norrie, even though our lives have taken different paths.)

As an 'expert' I was on £5 per week, paid straight into my bank account, which was a better wage than most people were getting at my age. A further advantage was that I had very few personal expenses because I travelled so much, so I started to put a little money aside.

I was a 'packet of rolling tobacco per week' smoker, apart from which I didn't spend much on myself because there was no call to do so.

Because I would be away for such lengthy periods, I knew nothing of the events of the past months (February/March) that would lead me finding my life's partner.

And it all started while I was away.

Mrs Cahill's Dance Band

Keith and his friends attended a place that was known as the 'Pavilion' that was frequented by an older group of people called the Belmore Social Set. (Belmore is a suburb near Campsie, and once a month they used to hold the 'set' dance.)

Keith's girlfriend was Norrie Weir whom he later married. Her parents were members of this set.

Most of the people were over-50 and everybody had to be well dressed. The men wore tails or dinner suits and the women wore long dresses and tiaras, if they had them. They specialised in the Pride of Erin, the Barn Dance, the Boston Two-Step and the other old time dances that were in fashion.

My future father-in-law, Ted Illingworth, played the saxophone and clarinet in various bands, one of which was Mrs Cahill's Dance Band, a four-piece band who played once a month for the Belmore Set at the Pavilion. The band comprised Mrs Cahill as band leader and piano player, a drummer, a violinist and Ted the saxophonist.

Patricia 'Pat' Illingworth was 17 at the time. She was prim and proper, tidy and industrious, and she liked music and dancing. One Saturday night when Pat's father was getting ready to go out to play at the set dance, he suggested that she join him, which she did.

So Pat went to the Pavilion, and sat quietly to the side of the stage watching the dancers, because she didn't have a partner. Everyone else had come with a partner.

Then Mrs Weir, Norrie's mother, spotted Pat by herself. She approached her and said, "Hello dear, are you here on your own tonight?"

Pat replied, "I'm here with my father."

"And who's that?" asked Mrs Weir.

"He's the saxophonist."

"Oh well, come down and meet the young ones," Norrie's mother insisted, and Pat was introduced to my friends – Keith, Norrie and another couple.

They then swapped partners a few times so that Pat could have a dance and everyone had a good night. Had the incident ended there, I probably would never have met Pat. However, for some reason, Norrie said, "If you come next month we'll bring a partner for you." They certainly didn't have me in mind, because at that time I wasn't even in town; they had picked out some other fellow. I was away in the country and the Smiths never exactly knew when I'd be back.

So they arranged for this other fellow to go on the blind date with Pat, but he backed off at the last minute which left them with the awkward predicament of desperately chasing around to find another dancing partner to fulfil their promise to Pat. I knew nothing of this, having come home after being up-country for four weeks.

Girl Without A Partner

During the 1940s, local town halls were popular dance venues, such as the Leichhardt, or the Marrickville, or the Botany, or the

Petersham Town Hall. They usually held a dance once a week. That night I had decided to go to the dance at Petersham Town Hall. Most of the time I would go out on my own.

I was dressed in my full three-piece suit and I was half way out the door when Keith and Norrie stopped me and said, "Don't go to Petersham; come with us to Belmore! You've got to come with us tonight."

I said, "Why?"

"Because there's a girl coming who hasn't got a partner," Norrie replied. "And I promised to bring a partner for her."

I said, "I'm not going on any blind date. Anyway I've made other arrangements!"

Eventually, after some considerable persuasion, I decided to go along with them. And that's the night I met Pat, and more than 50 years later I can still remember it very clearly.

We met under the grandstand at the Canterbury Rugby League Football Club's ground at Belmore on 15 April 1950.

All the other women were in evening dress. Pat wasn't. Pat was a 'modern girl'. She was wearing what was called the New Look. She wore a long, black skirt and a white blouse that her mother had made for her. We had a lovely evening together. I was smitten from the minute I met her

I was her blind date, and I've been blind ever since.

In the old days a man went up to a woman and he would ask if she would like to dance with him. But the night I met Pat was 'Ladies Night', which meant the ladies had to ask the man to be their partner, not the other way round. And Pat – being Pat – was never going to ask me even though I was accompanying her as her partner and we hadn't yet danced.

So in frustration, I said to her, "Would you like to ask me, if I'd like to ask you, if you'd like to have a dance?"

She said "yes".

Our first dance was the Jazz Waltz, of all things! I knew the Jazz Waltz so we got up and danced. Unlike the other dances, the Jazz Waltz didn't come out in the 1910s and 20s – it came out in the 1930s and 40s and it wasn't very popular with members of the set because it was a modern dance.

So we danced on our own in the middle of the floor.

Alone!

And all the old ladies sat around the floor were saying, "Who's this couple who are taking over the dance floor?" And, of course, her Dad was up there playing his saxophone –playing for us.

When the band stopped playing, I looked up and saw that we were right underneath a bunch of flowers, so I said, "We're right under the mistletoe, which deserves a kiss." This offended Pat's sense of propriety.

We didn't have that kiss, and I didn't call her for a few days, but we effectively started our lives together after that dance.

I met her father that night. We had a real affinity. We both liked old motor cars, we both liked music and we both loved Pat. The joy of Pat's father was his 1925 Dodge car. He lavished care on the old girl, not even letting me drive it.

However, a few things had to change for me to impress Pat. She liked my 'wavy hair', she liked the fact that I didn't drink or swear, and she felt at ease with me right from the start. But she thought I was a bit 'forward', she didn't like my dress sense, and she didn't like my motorbike – even though Ted thought it was a great bike!

I had the LE Velocette when I first met Pat; it was the first of the modern motor bikes. It had front and back-wheel suspension, telescopic springs and a little radiator, so it was water-cooled. You couldn't hear the engine when you were travelling along. It was the quietest motor bike on the road. It had sheet metal faring over the whole of the bike, so it was also smart looking. It is a popular exhibition item in Motor Cycle Museums today.

Nevertheless, Pat remained unimpressed, so I saved up and bought a car – a 1936 Austin Seven.

Nobody Else

I wasn't mad keen to have a girlfriend, although I had previously had 'girls as friends'. As for my relationships with women, I had only had my mother, my sister and my two half-sisters. As a result, I wasn't a woman's man. Although Pat was not my first girlfriend, she was my first *serious* girlfriend.

I can't imagine that anybody else would have been as good for me as Pat has been in every way. If I never saw anybody but Pat for the rest of my life, I wouldn't be unhappy. She taught me to stop being self-centred. She taught me to take a look at the rest of the world. She made me believe in other people.

As far as I am concerned, the greatest thing she did for me was to change me from the person I had been, to the person I became. Since my childhood, I had built a barrier between myself and other people because I had been hurt a number of times over the years. Being with

Pat was the first time I didn't need the barrier – and Pat did that for me.

Every time I got back from the country, Pat and I would get together. In those days a phone call from the country was quite a big deal and even though I wasn't much of a writer, I wrote to Pat almost every day while I was away. She's still got those letters tucked away somewhere.

Pat

Ted – Pat's father – and I got along well. He was born on 2 April and I was born on 2 April, which probably explains why we were a bit alike. He was a licensed plumber and a semi-professional musician. I was an unlicensed electrician and a lounge room musician. And as I said earlier, we both loved old cars.

Although Pat's parents – Ted and Marge – had a good marriage, there was also a deep sadness in their lives because Pat's brother Kevin had died from a leaking heart valve when he was nine. It hurt Pat's mother and father very badly.

The family lived in one of the oldest parts of Sydney – Banksmeadow, virtually on Botany Bay and named after the botanist, Joseph Banks. They lived in Wiggins Street, a short walking distance from Botany Bay and close to a massive industrial area. It incorporated the Mobil, Ampol, Caltex and BP terminals, the Johnson and Johnson complex, ICI Australia, Kellogg Australia and two miles of similar enterprises that occupied the land between their suburb and Maroubra. Beyond this, they were also ringed by golf courses – the Eastlakes Golf Club to the north and the Botany Golf Club to the south and on the bay. Across the water was Sydney's Kingsford Smith Airport.

Campsie, where I lived when I was in Sydney, was three suburbs away.

Pat was born 30 July 1932. She had attended the Banksmeadow Public School that virtually backed on to her street. She was a good student. She also went to Gardeners Road High School where she got her Intermediate Certificate before leaving school at 15.

Her first job was as a typist/clerk at Murray Brothers Timber in Waterloo; at that time it was a big timber company in Sydney. There Pat got involved with bookkeeping and she discovered that she had a natural aptitude for accountancy. When she was 16 or 17 Pat decided that she wanted to become a chartered accountant. Although that idea disappeared after we became 'serious', in a strange way our meeting and subsequent business partnership with Driza-Bone and other interests fulfilled her expectations beyond her wildest dreams.

Pat was smart girl in the work that she did for Murray's Timber. She had a great retentive memory. She was a learning-type person and she never stopped learning how to do the job. She was terribly meticulous. She wouldn't do part of the job; she'd do the whole job. She wouldn't shortcut the books. If the balance were out by only 6d, she would track it down. That attention to detail is one of the greatest assets any company can have, because proper financial management is imperative if a company is to prosper.

Pat worked for Murray Brothers, and – projecting into the future – she worked for PDF Food, H&H Timber, Nicholson Brothers & Lucas, Australian Timbers and eventually Usher & Guest, which she left when we started our own business. This led us to Driza-Bone.

Ted Illingworth

Pat and I met on 15 April 1950. We got engaged on 15 March 1951 (and we later married on 15 March 1952). I asked Pat first, and then I asked her parents. There was no way in the world I would have asked her second! I have a lot more respect for her than that!

There was no argument from her parents about us getting married. The only possible problem was she was a bit young. Pat was not a worldly person. She had lived at home and went to school, after which she lived at home and went to work. I was 24 she was 18, and consequently I was a bit more worldly-wise than Pat.

I was always very comfortable with her mother Marge as well as with her father Ted –

when he was alive.

However, October 1951 was a bad month. There was a drought, the beginning of the Communism scare – and a death in the family. Pat's father died exactly one week after his 25th wedding anniversary.

At that time the Renault car company had released a rear engine car called the Renault 8. It was quite similar to the Volkswagon Beetle, and Pat and I had convinced Ted to buy one. Ted had ordered the car, which was due to be delivered on 17 October. So –sentimentally – on the evening of the 16th ,Ted said to Marge, "Let's go out in the old girl (the Dodge) one last time and go see Hammos."

The 'Hammos' were the Hamiltons who lived in Paddington and were friends of the Illingworths. So Marge and Ted hopped in the 'old girl' and drove over there on a cold night and Ted parked in the backyard of the Hammos' unit, maybe a 12 or 18 inches from a fence.

When he came to drive it home, the car wouldn't start. It was quite common to get the crank out to start the car but, being trapped by the fence, he cranked it with great difficulty. And then he got furious with it because it wouldn't start.

By the time the car eventually got started, Ted had got himself into a bad state.

He got as far as Kensington, when he had a heart attack and had to pull up sharply by the side of the road.

Marge raced to a telephone box and rang Pat who was back home with me. We jumped into my Willys, drove over and picked them up. Pat drove my car back to Banksmeadow and I drove the Dodge.

In those days you didn't go straight to the Emergency Hospital Unit, you telephoned your local doctor first. We got home, put Ted to bed and rang the doctor who gave Ted an injection of some kind after which he seemed to pick up. The doctor said that we should look after him and that "he'll be right".

But Ted wasn't 'right'; he just kept on getting worse. We were with him that night when he died.

Naturally, Pat was very affected by this sad death.

Luckily, I was there to take over the responsibilities of what had to be done. We sold the old Dodge for £100 and of course cancelled the order for the Renault.

After Ted's death, Pat continued to live with her mother and I spent a lot of time at their home. I was one of the few people who could control Pat's mother. She needed a husband really. I resigned from Cooper Engineering because I was no longer happy travelling up-country for weeks at a time. I wanted to be where Pat was.

My father and his wife Elisha would sometimes visit – he liked Marge's cooking, and she liked him. She considered him to be a well-dressed, charming person, so I didn't spoil the impression by telling her that his Scarf Brothers suit was about to be repossessed.

The Selling Profession

Although, after watching my father work, I had mixed feelings about the selling profession, I needed work so I applied for a job as a salesman for a company that sold vacuum cleaners, refrigerators and small washing machines door to door. The sales manager was an odd-looking man called Ron Martin, a tall man with a 'wall' eye (which is an eye that wanders all over the place). It was he who convinced me that selling was a most reputable profession.

He explained that a salesman on commission could earn a better-than-average income. And so, with three other young hopefuls, I attended his sales training program before hitting the streets. After two weeks of lessons, the big day arrived – we went out to make our fortunes, but every door I knocked on was a *no*. The only thing I got on day one was a free cup of tea.

The second and third days were better in the sense that I at least got to demonstrate the product, but I still hadn't sold a thing. The other three fellows had made some small sales, but nobody had been particularly successful.

So Ron decided to demonstrate how to sell. He asked one of us to pick any Sydney suburb at random, and the answer was Bexley – which comprised middle-income earners. Then a second salesman chose a street within Bexley, at random. The house number was likewise chosen at random, and it was decided that the appliance to be sold was a small washing machine. The next day the four of us accompanied Ron to that address to watch the master salesman at work.

His first step was to get in the door, which he did surprisingly quickly. His next step was to gain permission from the elderly couple to demonstrate the machine, which took another five minutes. He then convinced the pensioner couple that they really needed the machine, particularly as they were getting on in years. They agreed and they paid cash, even though they had previously said that they had no money whatsoever and were poor old pensioners.

I never forgot this important lesson – the salesman's creed: *create a want and then supply it.* Ron also went on to explain that these people did have a genuine need for the machine, and that he would never have lied to them to wrap up the sale.

Later in life I earned good money as a salesman. Most of my businesses relied on salesmanship, particularly in the days when we were selling safety equipment. Later in life when we had Driza-Bone, we had to create a need and then supply it before we could take it to the whole world.

CHAPTER 7

Mr and Mrs ~ Driza-Bone Comes Later!

I was a salesman for some months before Pat and I got married. Though I did not stick with door-to-door selling, my sales training under Ron Martin served me well in my next job, which was selling cake to shops from a van.

The Kingsford Cake Company that I was working for had their manufacturing premises just down the street from where Pat lived. Their locality was one reason why I wanted to take the job on because they were conveniently close to Pat.

I was one of the first cake suppliers to Blacktown, which was then an outlying suburb between Parramatta and Penrith in the days when Sydney's western suburbs were a semi-rural area comprising small chicken farms, strawberry farms and other small acreage farms. Today, of course, Parramatta-Blacktown is the geographic centre of Sydney.

In those days, Sydney retailing comprised mixed businesses, which were mostly located in suburban shopping strips. These strips had a butcher, a milk bar, a fruit and vegie shop, maybe a haberdashery shop, and so on. People couldn't buy fresh cake locally, particularly in the outer suburbs. It was an era of the mixed business rather than the specialist store. Most of the retail shops got their supplies from sales vans.

The Kingsford Cake Company had been started by Roy Glover, a former employee of a rival called Kenso Cakes. Kenso's Cakes and Kingsford Cakes were the two main suppliers to shops. Later, another firm started up, called Gartrell White. It still exists today, as Tip Top Bakeries.

As far as delivering the cake, today you wouldn't be allowed to do what we did in those days. We used to carry the cake from the van to the shop in our arms on greaseproof paper. We'd walk in and put it into the display case, which we hoped the shopkeeper had cleaned before our arrival. This was not necessarily the case. A customer would then come in to purchase a piece of cake and the shopkeeper would cut off a slice, weigh it, put it into a brown paper bag - and that's how it was sold.

Mondays and Thursdays I went to the eastern suburbs as far as Dover Heights and Vaucluse, through to Matraville, back up to Mascot, Alexandria and back into Botany. Tuesdays and Fridays I went west as far out as Blacktown, Seven Hills and places like that.

Wednesdays I would drive over the Sydney Harbour Bridge, up the North Shore, along the Pacific Highway. Again, there were no specialist cake shops in those little shopping strips that lined the highway between North Sydney and Hornsby.

The Philishave

I had previously lived all my life out of suitcases – from Peak Hill to Sydney to Murwillumbah, and everywhere in between. I had never settled in any single place for long. Furthermore, from the age of four I wasn't used to being answerable to anybody but myself – not to my mother, my sister, my father, my uncles and aunts, not even to my employers. I had lived a pretty independent life.

Therefore, I needed to make some adjustments before I could reach the good understanding that Pat and I have today.

We don't have many arguments, although we have had a number over the years. But none of them has been serious and none of them has lasted longer than 10 minutes, so we're very lucky that way.

Our very first argument was about time.

I have always been a great believer in being on time. If I make an appointment with someone, I get annoyed if I am late.

One night Pat and I had a date to go to a stage show which started at 8.00pm in the city. When I turned up to pick Pat up at 7 o'clock, she hadn't begun to get ready! She hadn't even had her shower - yet we had to be in town by 8 o'clock and there was no way in the world we were going to make it on time.

In those days we travelled everywhere by tram. They weren't terribly reliable and they were fairly slow. Banksmeadow was on the tramline and – apart from the wait – it was a half-an-hour's trip to the city, after which we had to get to the theatre and find our seats.

I had spent a lot of money buying those tickets, so I lost my cool and told her a few home truths about what I thought being 'on time' meant and she was very upset with me. Wow, was she ever!

That was our first big argument. Since then, we've both become strict about time management, and she's become as fanatical as I am.

Another argument worked the other way around and taught me how insensitive I was to other people's feelings. It came about this way: because I had sensitive skin, I used to shave only once every second day and I had this sharp growth. Pat and I would kiss, and I would scratch her face, so she finished up getting a rash. This particularly happened after I'd been away four or five weeks, because when I'd come back our kissing sessions were probably a bit more intense.

So (in self-defence) Pat bought me a Philishave shaver with two rotating heads. They were a vast improvement on any other electric shaver and when she gave it to me I was apparently not very grateful. I shrugged my shoulders, said "Oh thanks", I put it down and didn't think about it.

Pat had spent a lot of money buying this shaver. They weren't cheap – she was only earning a few pounds a week and she had probably spent more than a week's wages on the dashed thing.

It wasn't our first fight, but it was the first time that Pat had explained to me what I should be like, and the first time I became aware of what type of person I really was. I was self-centred. I had never taken an account of myself. I had never looked at Frank Fisher before and thought, "Hey, you're not always a nice person."

I never cared what other people thought of me. I was happy with me and I didn't need other people. I didn't care about other people's feelings. I was content to sit by myself and I felt quite comfortable in my own company.

Now, for the first time in my life, I *did* care about what someone else thought, and Pat made me aware that other people were entitled to be given some respect and entitled to a proper 'thank you'. I hope I learned that lesson well because I wanted to please this wonderful woman.

Married

Pat and I got married on 15 March 1952 at St Matthew's Church of England, Botany. It was a big wedding for the time – with nearly 100 guests. My boss, Roy Glover, was among the guests. Although Pat is not a churchgoer, she believes in God a lot more than I do which is one reason why we had a regular church wedding.

My best man was Keith Smith, and my groomsman was Teddy Angus who was married to Pat's best friend, Pam, her matron of honour. Pat wore a crinoline floor length gown, a Mary Stuart head-dress and the bridesmaids wore orchid pink. The songs included *O Promise Me* and *Because.* It was a nice wedding.

Accompanying my father was my new stepmother, Elisha. Her background was from a well-to-do New Zealand family who had made their fortune in dairy farming. My father told us that she was related to Sirs and Lords, but Shirley and I could never tell what was true as far as my father's stories were concerned. However, there might have been some truth in it because she had enough capital to enable my father to live off her for the last years of his life.

If he hadn't married her, he would have probably finished up in a Matthew Talbot Home or some such institution.

My father could charm anybody; he certainly did so at our wedding. Well dressed and well spoken, he described Pat as his 'glorious' daughter-in-law of which he was 'very proud'.

He was quite a hit with the guests until he made a mess of opening the wooden beer keg and accidentally sprayed a nearby baby with beer.

I guess when we go back to those times, a numerologist could have fun with the number 15. I met Pat on the 15th, we got engaged on the 15th, we got married on the 15th. Later, our son Stephen was to have been born on the 15th (July) - but he hung around for three extra days and came the 18th.

Furthermore, we are godparents to our friends' daughter Lindy, who was born on our wedding anniversary, again the 15th.

I even started this book on the 15th – coincidentally, on my wedding anniversary.

And So – To Wed!

Pat and I spent our first night at the Wentworth Hotel, which was supposedly one of the best hotels in Sydney at the time, but a great disappointment to Pat and I because it was old, decrepit, tired and worn out. Shortly afterwards it was pulled down and rebuilt into the edifice we know today.

We went north the following day, in our 1936 Willys car. I converted the back seat so it lay flat – and that was our bed. We slept in the car most nights. There were no such things as caravan parks in those days, though every town had a little park with toilets and taps. We'd wash ourselves with a hand cloth. We had a dish and a billy to make hot water on our Primus stove and we'd also have a shower and pour water all over ourselves. Pat and I were so modest in those days we'd wear a swimming costume when we did that. Then every three or four days we'd stay in a hotel to have a proper shower.

We drove to Brisbane first via Wauchope, where we stayed a few days with my relatives – the Currans, the Kellys, and the Doyles – and we then came back through country New South Wales to Dubbo, and from Dubbo back to Sydney.

Pat had not been out of Sydney before our honeymoon. She had never been further west than Katoomba or further south than Wollongong. It was also the first time she had been with me without her Mum around, so this was quite an adventure for her.

Travelling west, we were in the heart of Driza-Bone country. It was the land from which Richard Mills had spawned the Patersons, the Youngs and the Fishers. It was the land I had worked when I was a child, and to which I had so often returned as a teenager and young adult.

It was where I had learned to read and write, where I had learned about the 10 Commandments, where I learned to drive, where I had learned about motors, and where I had learned to shear, plough, hoe and fend for myself. It was the land I had looked down on from a joyride in a Battle of Britain Hurricane fighter plane.

It was also where I had returned as an 'expert' for Cooper's Engineering.

I introduced Pat to Big Auntie, Uncle Rowley and a lot of other relatives. I showed her Tomingley, Eumungerie, Parkes, Peak Hill and the humpies at the Dubbo Flats.

It was such a far cry from Sydney's Botany Bay that Pat was not impressed. She had no idea that the western plains were so dry. She was disgusted that wheat sheds bred rats. And when we were there in 1952 the farmers on the western plains were fighting back a rabbit plague, and the effects of myxomatosis was an unpleasant sight.

Married Life

When we returned from our honeymoon, our first home was with my mother-in-law at Wiggins Street Banksmeadow. This was not ideal for us because, as newly-weds, we needed a space of our own.

Pat's mother's house was a brick Federation cottage. It was a double-fronted house on a narrow block. We often laugh that Pat lived in a brick house before we were married. It doesn't sound terribly significant, but there weren't many people we knew who lived in brick houses, they mostly lived in weatherboard or fibro homes.

We walked down the side to get to the front door. It had a lounge room, a bedroom and a tiny verandah in front of that. Behind that was the second bedroom which was 12' x 12'. The main bedroom was 12' x 14', with a bathroom with a shower between the two rooms. Behind that was a family room about 10' x 10' which had a fold up bed where I used to sleep before we were married. At the back was a verandah that went across the width of the house. At one end was a laundry and at the other end was a kitchen. There was an inside toilet, which Ted – as a qualified plumber – had put in.

At the front of the house, next to the main bedroom, was the lounge room, which had the piano in it. This was Pat's piano and the first thing she had ever bought. Our son Stephen still has it.

Marge was a good cook, which made it pleasant for all of us when we'd come home from work.

Being a good cook gave my father and Elisha a reason to visit. They lived in a walk-up unit in Victoria Road, Potts Point. It was a bed-sitting room with a little kitchen and a bathroom down the hall. The main room itself couldn't have been more than 15 x 20 feet with a little balcony off the front verandah. It wasn't anything exciting, even in those days.

After we were married my father and Elisha would frequently turn up at Marge's place – mostly at weekends – at meal times. They no sooner finished the meal and Elisha would say, "C'mon now, we've got to go to so-and-so's place" and they wouldn't even help with the washing up. In those days – before dishwashers – it was customary for everybody in the family to help wash up. However, even without my father's mealtime visits, the relationship between Marge, Pat and I was beginning to show signs of wear and tear. We wanted a place of our own – we *needed* a place of our own – but we didn't feel right about leaving Marge all alone.

However, after some time Marge began going out with Wally Hodder, a neighbour, and it was a real love match. Wally was a true gentleman and a gentle man. She eventually married him, after which we felt comfortable about moving out.

Boom and Busts

I don't know why everybody who writes about the 1950s talks about 'the post-War boom' as if it were a period of continuous national prosperity. In reality, we had the boom and bust days. They seem to forget that in 1952 we had a period of economic recession. A lot of people didn't have jobs. It wasn't as easy to get a job then as people now make out. Money was very tight. We paid a third of our income in rent, a third of our income in food and a third of our income for living expenses.

Ups and downs were the way of things in the days when the unions got involved. The Menzies years, when Robert Menzies became a long-term Prime Minister of Australia, began in 1949 and seemed to last forever. Certainly there was low inflation under Menzies – maybe 1%-2% – in the days when we thought 0.5%-1% was high!

Pat and I were aware of the political situation but it didn't have much influence on us. We certainly didn't get involved in the politics of the time. We never even thought we could actually attend a political function, for that matter.

It didn't really make much sense when Menzies banned the Communist Party of Australia and launched his 'Reds under the bed' paranoia. We had just come out of a war, so we had an understanding of real war. What we couldn't comprehend was the idea of a 'cold war'. We were either fighting a war or we weren't, so I never figured out what the Cold War was.

During this period a new mood swept the country. It began around the time of the coronation of Queen Elizabeth II. Like most Australians we knew, we didn't care very much one way or the other about the Royal Family, although when the Queen made her royal tour of the 'colonies' shortly afterwards, Pat was very happy to go to the races 'with the Queen'. This came about when she worked for Australian Cotton in O'Riordan Street, Alexandria, whose owners were very much a part of the establishment. They played polo, they went to the races and they were involved in all that sort of stuff.

The company was in receivership and the receiver (who was a MLA) had two tickets to attend the Randwick Races, which the Queen was attending. He couldn't go, so he offered the tickets to Pat and one other office girl. They got dressed up and sat – not quite in the Royal Box – but almost. It was something Pat enjoyed.

In the years before television, Pat and I would listen to radio shows like *When A Girl Marries, Wyatt Earp* and a show about a space ship that flew to Mars. Sunday nights it was always the *Lux Radio Show*, which featured an hour-and-a-half play that everybody would listen to. We enjoyed all those sort of things.

We lived a full life. We didn't just sit at home and twiddle our thumbs. Pat used to play basketball once a week, every so often we would go out together and play tennis, and we'd go to the pictures once a week. We would also go to dances. We read whatever we could get our hands on. Books were expensive – which is the reason why we never bought hard covers; we always bought paperbacks.

Strivers

Although W C Fields was never my favourite comedian, he said something that I have never forgotten: "Live above your means, then achieve it."

In a way, Pat and I have always lived by this maxim, always reaching out to improve whatever situation we were in. I'd put it this way: "Don't be too satisfied with what you've got – go for something higher, know you can do it, then make it happen."

Pat and I were always looking for something better. We were both strivers. We wanted to do something special with our lives.

It's very hard to explain those feelings.

We never were satisfied to accept the status quo; we always wanted to improve on it. We were always moving, to bigger and better things.

Our first home was a 10' x 20' garage.

Our next was eight squares.

Our third was 15 squares.

Our fourth was 23 squares plus.

And so forth...

Pat and I were always hard workers and we never shortchanged our employers. We were prepared to give a full day's work for a full day's pay. We were the early birds – the first to arrive in the morning and the last to leave at night.

This was a habit we did not break when we started our own business in 1962.

Part 2

Protective Safety Clothing

8. Safety Salesman

9. Superior Safety Salesman

10. Our Own Business

11. Industrial Safety Equipment

CHAPTER 8

Safety Salesman

In 1952 I continued as a van salesman, although now that I was married I was anxious to improve my lot. I was sick of selling cakes and there certainly wasn't much future in it.

I had been working for the Kingsford Cake Company for more than a year. I couldn't go any further with them. Sure, I might have applied for a promotion to sales manager or something like that, but whom would I be kidding? The sales manager of one van wouldn't have been a big deal.

I don't know why I happened to pick up a *Sydney Morning Herald* on a Thursday, but I did. And I looked up the Positions Vacant section where I saw advertised: "Wanted: safety salesman to sell industrial safety equipment to industry." This sounded a lot more interesting than selling cakes.

I wondered what 'industrial safety' was, because it really didn't exist in Australia as an industry at the time. For some reason I had a feeling that I wanted to get involved. I had certainly seen many workplace accidents that were unnecessary, so I responded to that advertisement.

Ron Millar

I rang and made an appointment to meet Ron Millar, the sales manager of Nicholson Brothers & Lucas Pty Ltd, which had their office in Surry Hills near Central Railway Station, in Sydney.

Ron told me that Nicholson Brothers & Lucas was an expanding business. They had established themselves primarily by making industrial leather gloves and had recently decided to expand. They wanted to employ four new salesmen – one for Newcastle, one for Wollongong and two for Sydney. He interviewed me, along with the other applicants. I rang him the following day to find out if I was successful. I rang him four or five more times and he said, "If you're this keen about the job, maybe you should come down and have another interview." And so I saw him again the following day and he gave me the job. He paid me a retainer of about £5 a week, plus 5% commission on sales, and car allowance. I spent the next three years working for Nicholson Brothers & Lucas.

Ron was about 5 ft 7 in tall, in his late-30s or early-40s. He had a long, thin face and a big smile. Ron taught me a lot, and in time, he had a big influence on my attitudes to sales, work and business. Before working at Nicholson Brothers & Lucas, Ron had been a car salesman selling Standard Vanguard motor cars, which were made in England.

Ron was an enthusiastic man; everything about him was about *today*! When you were around Ron, things would always be happening. It was always, "Do this! Do that!" *Bang bang!*

Ron employed me and Dick Tovey for the Sydney area, Teddy Mitchell for Newcastle and another salesman for Wollongong, who we did not see very often because he lived and worked down there.

The managing director of Nicholson Brothers & Lucas, Clive Nicholson, had worked for BHP at Port Kembla. He started the business mainly to make industrial safety equipment for BHP and the Big 7.

The Path To Driza-Bone

And so I began my new career – in fact, the path that would lead me to Driza-Bone – on 15 June 1952. (There's that '15th' again, precisely three months after Pat and I had been married.)

The four new reps started work on the same day. Ron gave us two weeks' training, teaching us what industrial safety equipment was all about. Ron showed us the gloves and explained in detail why they were made in a particular way. He showed us the various types of goggles, explaining which should be worn for certain jobs, and he went right through the Nicholson Brothers & Lucas range, explaining each product. He also gave us a stack of American and English catalogues which he expected us to read and thoroughly understand. Then he took us to the Sydney factory at Guildford and showed us how the gloves were made. We also saw other protective clothing such as aprons, leggings, spats and welding jackets, being made and (of all things) we also saw them making sponge cushions for some reason or another.

At last, we were designated our respective areas, and mine began at Redfern and then south to Alexandria, Mascot, Botany, Matraville, the south-western industrial areas and beyond. One reason this was to be my territory was because Pat and I had lived in Banksmeadow. It was the main industrial manufacturing part of Sydney at that time, so I was given the most important Sydney area to handle, of which I was very proud. I was determined to do a good job.

I was not without some sales experience. I had watched my father at work, I had been trained in door-to-door selling, and my time as a van salesman also involved a sales pitch.

First Calls

Pat was very pleased for me in my new job; somehow we must have both sensed that something was afoot and that this was the beginning of the rest of my life. I was very nervous when I got up the first morning and dressed in a full three-piece suit and tie. She could see that making this job a success meant a lot to me.

And away I went!

We were not given a list of company names as contacts; we had to cold canvass. For my first sales call I chose the first company on the right-hand side of O'Riordan Street, Alexandria, a long road which runs virtually parallel to Botany Road and was in an industrial area full of business activity. I had sold the Willys, and I drove there in my 1947 Chev Fleetwood. It was a nice big four-door car. Clive Nicholson paid me £6/12/0 a month to run the car irrespective of the miles I did. It certainly wasn't enough money to pay all my car-running costs, because by this time we had moved from my mother-in-law's place and we were renting a flat in Miranda. This move added to our fuel costs in this early period, because I had to travel back and forth to report to the head office at the start of every morning and at the end of every afternoon. (Salesmen had to pack their own orders so that the orders would go out the next day.)

And so it began. My very first call was a firm called Concrete & Terrazzo, makers of terrazzo and concrete steps.

I took a deep breath, crossed my fingers, opened the door, walked in, handed my business card to the receptionist and said, "Hi, I'm Frank Fisher from Nicholson Brothers & Lucas, industrial safety equipment." And then I asked to speak to the person who handled the buying of their industrial safety equipment. From then onwards I had no idea what was going to happen. Hopefully, someone was going to come out and talk to me; if not, I would be on to my second call within the next 10 minutes.

Then an Italian man came into the reception area and talked to me about First Aid equipment. I explained that industrial safety equipment was not First Aid, but something that *prevented* accidents rather than repair the damage. Eventually we got around to talking about gloves and he told me he used White Cross rubber gloves that were imported from England, which he admitted were hard to get as they were in short supply. The White Cross was a rubber glove that was very popular with the concrete and terrazzo industry.

"We sell a similar product; perhaps we can supply them to your firm?" I responded.

I could see that I had captured his interest, so I kept on talking. I said, "I haven't got a price with me right now, but if you give me your order I'll let you know the price at my soonest opportunity." I was doing the sales pitch without even knowing for sure whether or not we stocked it, but I took his order anyway.

I went back to the office that afternoon and found that we did indeed stock a competing product.

First sale, first strike.

My next order came from further down the same road at Australian Cotton, where Pat worked. I sold them 500 nose and mouth respirators similar to the type doctors used.

Australian Cotton needed them because cotton fluffs up. It's quite dangerous to inhale the fluff, because it can cause lung problems, so the employees had to wear masks. I also got their order on day one. That was my first day on the road and I had made two orders. The value of the two of them didn't amount to much in dollar terms, though it certainly lifted my morale, making me realise straight away that when people have a need, a salesperson can fill it.

Later, I had to go back to Concrete and Terrazzo with the price of the gloves, I got their confirmation order and I had to make arrangements to have them delivered. It was a bit of a hassle because we didn't actually carry them in stock. We had to get them from a firm in England called V-Dip, whose agent was in Australia. Nevertheless, a sale was sale!

In those first few days I decided to call on every single company in O'Riorden Street, including a big company called Austral Bronze which was next door to Concrete and Terrazzo.

I went in and asked to see their buyer and got quite a surprise when he said, "Oh yes, we already buy from you." Their managing director was Clive Nicholson's friend. Nicholson Brothers & Lucas wasn't going to give me this business (nor any commission). In fact, they later instructed me not to bother to call on them, because it was a 'house account'. But I went in anyway. If a door was there, I knocked on it.

I really enjoyed selling safety equipment; in fact, there was only one thing I found genuinely frustrating about the whole experience. Over the next two-and-a-half years my commission rate went down to zero. In the early days it didn't matter because none of us was earning much. Our sales were small and 5% of next-to-nothing was still nothing. Then I started to get good at selling. As soon as I started to increase my take-home pay, Clive Nicholson would either reduce my commission or increase my target.

I was furious when he did that, so I worked even harder and got another bonus the next month.

In retrospect, there were many stupid aspects to their sales policies because something as significant as safety cannot be sold on Let-Me-Do-A-Deal-For-You principles. It's more like selling insurance, because the buyer is ensuring that their employees will not be injured. You've got to first convince the buyer that safety is good value.

You have to explain that every injury costs money to the company. The person who gets injured is traumatised, plus that person's workmates are probably horrified and certainly de-motivated. Add to that the downtime costs and the cost of breakages and it is very apparent that accidents are a very significant cost factor. They say the loss ratio is 5:1 – that is, if a person loses one day's wages through injury, the final cost to the company is five days' wages. If the employee loses one month's wages, the total cost to the company is five times that.

Good Salesman

Clive Nicholson taught me a lot of things. In many ways he was a very nice person – even though he was a smarty as far as his commission and car allowance policies were concerned.

I'm don't mean to be a skite, but I was a good salesman because I had learned the trade, and I still am a very good salesman. Dick Tovey and Teddy Mitchell were never as good at selling as I was, not because they didn't work as hard but because they didn't have my background. The variety of jobs I'd held in the past stood me in wonderful stead.

By this time, I had worked at least 46 jobs so I had a fair knowledge of what people needed to do their jobs, whether they were working with motors, industrial equipment or machinery, and I also knew how most factories operated. (For example, on the farms, I used to grind the shearers' blades and combs, so I understood how to grind things.) In my sales capacity, the people I was dealing with were the people on the floor – they were the factory managers. I could talk to the foreman or leading hand at their level. It mightn't sound much, but if I could talk about grinding to someone who did it all day, I would get a more sympathetic hearing than a novice salesperson who might start by asking, "What is grinding all about?"

So when I walked in and said, "I'm here to sell industrial safety equipment", if the buyer said, "My main problem is dust" I understood that he meant he had a problem with factory dust created by grinding, polishing or something of that nature. Anyone working in such an environment needs to wear a proper respirator. I understood when they polished something that had been plated, the toxic dust would get into the air, and a flimsy mask or a handkerchief over their mouth was insufficient protection.

I was also aware of what the factories looked like inside, and when I was talking to the buyer about safety equipment, he and I would be talking the same language. That gave me a great advantage over our other salesmen because they had to learn about factories, as well as learning about safety equipment, from scratch.

Government Attention

I had a natural affinity for safety and all that went with it. I also liked the results that I was getting because there were times when I was told that the equipment I had sold had saved somebody an injury. Other times I was told, "We don't have that problem any more because last time you called you sold us the equipment that reduced those injuries." I was pleased to be working in an industry that had a significant effect on people's lives, their injury levels and even their earning capacity. It gave me a lot of personal satisfaction.

It may be difficult to believe, but safety equipment in those days had no official standards. In 1960 I sold the first approved safety helmets ever sold in Australia to a Sydney firm called Hutchinson Brothers Contractors. Industrial safety was becoming a very big issue at that time and the Australian Standards Association was coming into being. Everybody was talking about it, because the cost of injuries was horrendous. I was a foundation member of the Australian Standards Association.

By the early 1960s the respective Standards Authority bodies were beginning to set some ground rules for minimum standards at both State and Federal levels. Although Australia had two main manufacturers, Nu-Plac in Melbourne and Protector Safety in Sydney, most of the hard safety equipment was imported from America or England. (Soft safety equipment, such as leather, plastic and rubber gloves and aprons, was made in Australia.)

A lot of wonderful people worked very hard in those years to try to get safety standards into the workplace. In New South Wales safety was under the jurisdiction of the Department of Labour and Industry.

No Safety Standards

As I said, before 1960 there were no standards for safety helmets in Australia. Building construction was going at a great rate and a lot of people were getting injured from things dropping on them from even 10 or 12 feet. A hammer dropped on somebody's head from that height is enough to kill them. Half a brick falling out of a high wall could also kill someone. Yet in those days it was not customary to wear protective headwear on building sites. Only miners wore helmets as a matter of course. The most that builders would wear might be a felt hat, like an Akubra. Most of the builders reckoned that was as good as they could get.

Furthermore, the only helmets that were being used were imported. The best ones were the American aluminium helmets, which were quite good as far as absorbing impact, but a problem arose because there was only about half-an-inch space between the helmet and the person's head. A sharp object will strike through that half an inch and such a blow will still impact on the person's head, which could again be enough to kill that worker. As a matter of urgency one of the things that needed adjusting was the space between the harness and the wearer's head.

The other imported helmets came from a firm called Cromwell, named after the English town where they were made. They were made out of compressed cardboard, which is the same sort of stuff used to make school cases and suitcases. For heaven's sake, they offered little more protection than a cornflakes packet!

In those early days the safety equipment that was around was terrible. Imagine wearing a metal helmet when working on electricity wires? You'd get a boiled head!

Cliffy

The New South Wales Department of Labour and Industry built its policies on those overseas standards. The man who ran the safety side of the department in Sydney was Cliff Gartrell. Everybody knew him as Cliffy. He was a little man, he wouldn't have been more than 5 ft 6 in tall and he was very slightly built. Being well into his 50s, he seemed quite elderly to me.

Cliffy was a totally committed bureaucrat who had worked for the Department of Labour and Industry all his life. The head office for the department was in a four-storey building in George Street North, in the Rocks. The four stories were significant, because that was the height from which he and I used to use to drop a metal ball onto a helmet to find if it could stand the 40-foot pound test.

I was one of the first members of the New South Wales Australian Safety Council, which began in the mid-1950s. Cliffy and I got safety helmets from all over the world and we would put them through the 'four-storey test'. This was a very significant part of selling safety, because if I could get Cliffy to approve the headwear, I could then approach the industry and legitimately claim government authorisation. One of the first things Cliffy and I did was to increase the space between the hardhat and the head from half an inch to two inches (5 cm) of space between the top of the head and the inside of the hat. So when the helmet was hit, the blow would not impact on the head.

Goggle-Eyed

When I first started selling, safety goggles they were made out of aluminium and the lens were made out of window glass. People just didn't realise how dangerous those sorts of things were. A decent-sized blow would hit these glasses, break them and send glass into the wearer's eyes! I regret to admit that's what the very first safety goggles were like.

So the first thing we did was put a plastic lens behind the glass, which gave some protection from the shattered glass because at least the plastic didn't break. That was one of the very first changes we made from the goggles that were being produced from designs still using Second World War technology!

A welder might make his own welding goggles from the bottom of two blue bottles. In those days you could buy blue bottles, mostly used for castor oil and things like that. The blue would cut out the ultra-violet rays, no problems, but a welder also needs to insulate the eyes from the infra-red rays which will burn the cornea. Nowadays, of course, you've got enough protection from both problems of radiation, which they didn't have in those days.

These were quite significant changes that had to be made and there were lots of people innovating and creating new products. The market was wide open.

Most of our ideas came from an American firm called Wilsons. They were very big in industrial safety equipment and they supplied equipment to us. A firm in York Street Sydney acted as their distributor. They had a few contracts but they sat back and rested on their laurels, which left room for a Melbourne firm called Nu-Plac that started manufacturing a lot of copies of the Wilson equipment. They made quite good copies too.

Factory Accidents

I can recount many stories of industrial safety where we in the sector genuinely saved people from injury, and in some instances we saved lives. Selling safety was morally very rewarding.

I used to give lectures and conduct surveys on safety. I like dealing with people. I was always comfortable talking to people about safety. It didn't matter to me whether I was talking to the managing director of BHP or the sweeper in a factory in the back blocks of nowhere land.

A sheet metal company located at Rosebery had an injury rate of 22% before they started dealing with us. Imagine how much each one of their injuries cost the company – on a ratio of 5:1 - apart from the pain, the agony and the trauma?

When the company moved to an outer western suburb, and started re-thinking their safety arrangements, I was invited to be a part of their consulting team. When we got the new plant properly equipped to meet safety standards, the injury rate went down to .02%. I stress – from 22% to .02%. Safety equipment was a relatively small investment for such a company and everybody realised that it was good value.

Another example was a company called Hyster Forklifts at Milperra. I talked them into getting their staff to wear safety footwear. On the day I delivered the footwear, I was barely out the door before I was called back because an employee who was wearing the new safety shoes had dropped a metal shaft on his toecap. Although it had flattened the shoe, it had not damaged his foot. The manager told me that when the same thing had previously happened to another employee, he had lost a toe. The manager (whose name was Mr Challenger) called me back to complain – jokingly and audaciously – that they had only got one day's use out of the footwear! How cheap was that, compared to an injury?

Another factory I called on manufactured ceiling bats. This factory had an oil-burning furnace and a conveyor belt with no safety switch close by. I could see they faced a major problem if the furnace blew back and sprayed the burning oil on to the conveyor belt, because the employees would have to run some 50 feet to turn the boiler off at the switch.

Three days before it happened, I told the manager it could burn down. And it burned down for the very reasons I told them. That was a very expensive lesson, all because of a badly positioned switch.

I also convinced the manager at Otis Elevators into getting his staff to wear safety spectacles because a person sweeping the floor requires different vision to a person writing at a desk. Otis Elevators had cleaners who couldn't see as far as their broom heads because, for some reason, everybody in the cleaning department was over 50 years old. (Once you reach 40 many people's eyesight deteriorates to the stage where they need glasses, therefore all these 50-year-olds couldn't see what they were sweeping.) When I explained the problem to the manager he issued safety specs to the cleaners which they still wear to this day. The spectacles gave them the long sight needed for them to see the end of their brooms, and suddenly the factory was clean! And a clean factory is more productive than a dirty factory.

I've even got a letter from the managing director of British Leyland thanking me for my contribution to his safety program at the Zetland branch when they were in Australia. He said how appreciative he was for the contribution I had made to the company by supplying them with industrial safety equipment over many years. It gives you a nice feeling when the managing director of such a big company takes the time to say thanks.

Clive Nicholson

It became apparent that Clive Nicholson's attempt to set up premises in Surry Hills was an unfortunate move. There was no parking, it was too small, and for many other reasons it was an inappropriate location, so they relocated the store to premises in Cannon Street, Petersham, almost on the Parramatta Road.

Clive took over the premises from another company with which he had had some relationship, and he finished up buying the building. Because a lot of renovations needed doing, we salesmen came back after a hard day's selling and we worked our backsides off every afternoon, turning this plastics roofing operation into offices.

In the meantime, after I complained about the cost of owning a car, Clive asked me what kind of car I was driving. I was still driving the Chevrolet which I couldn't afford to run any longer because the money he was paying me per month wasn't enough to pay off the car as well as its running costs. I almost went broke keeping that car on the road. With the cost of petrol between 1/6 to 2/- a gallon, it was quite common for me to drive down to Wollongong, pick up a load of gloves and bring them back to Sydney. All this out of the meagre £6/12/- car allowance he had been paying me per month. Furthermore, it was hard to keep the Chevy on the road and I had to do a lot of work on it myself to keep it running. It had previously been a taxi and had been around the clock a couple of times.

So, after an unsympathetic hearing from Clive, we decided to sell the Chevy and we bought a 1952 FJ Holden instead – because it was much cheaper to run. Boy, that was quite a car. It was lighter, nice to drive, and a *Holden*!

That was important.

Unfortunately, I didn't keep that for very long because, on only £6/12/- a month, I couldn't afford that either. So we sold it we then bought a small Renault.

Pat Enters The Safety Sector

Ron Millar and Clive Nichsolson were well aware of Pat's ability as a bookkeeper because she'd called in a few times to see me at work when I was working back doing the packing, which we had to do every day.

Pat assisted me with my paperwork while I was doing the packing. After a while Clive noticed that she had the abilities that the company desperately needed, so he offered her a job. On her first day she was given her own office with a desk piled high with unfulfilled orders. Pat spent some considerable time getting them into sequence and then she proceeded to match the production schedules to the orders, and suddenly – out of the blue – the orders that I had been waiting to deliver all found their way to the customers. I was therefore given credit from this sudden surge of outgoing goods and I accrued a large bonus. And then, of course, my figures for the month went up and up – from £5000, to £10,000 and £20,000.

Clive Nicholson was quite upset that he had to pay me all this extra money, so he reduced my commission from 5% to 2.5%. As I continued to generate more sales, my commission went down to 2.5%, then to 1%, and finally to zero.

The salesman who had been assigned to the Wollongong area had not been a success at all, whereas I was so successful that I finished up adding his territory to mine. I used to go down there once a month and sell to all Clive Nicholson's house account customers. I would stay in a hotel for four days and I played a significant part in the supply operation. Although I was only known as Clive's representative, I effectively took over his mantle. In time I got to know all the safety officers as well as all the purchasing officers of all our client companies.

Then Clive bought a lovely big house with a water frontage in Sydney near the Gladesville Bridge. He simply transferred our role from salesmen into renovators, and he had us working there like navvies. He also got a couple of the other boys from his factory to work on his house. He got Ted Mitchell, Dick Tovey and myself to help him move. Dick and Teddy mowed his lawns and did other odd jobs and I even re-wired his house. For some time afterwards, one of us had to mow his lawns once a fortnight.

This wasn't what I had bargained for after proudly selling safety equipment for nearly three years!

Resignation 1954

In the end I don't know whether Clive Nicholson ended up being all that rich, but he probably had a reasonably good income. He was a fairly energetic type of person. He was quite authoritarian and he and I argued all the time, although in a sense we were still good friends. I was doing his sales job better than he had done it, and I think that probably upset him a bit.

By this time Pat was pregnant, so she left Nicholson Brothers & Lucas because she was having a pretty rough time with her pregnancy. She almost lost Stephen a few times during that period. She quit first, and then I left soon afterwards.

Clive Nicholson's way of paying me really sucked us dry. He wouldn't accept my resignation and always came up with reasons why I shouldn't leave. I gave my notice to him quite often and one day he simply accepted it. At the time I thought I had made a mistake by resigning, but it was one of those things I'd do just about every second week — and in the end, it was over.

In the meantime Pat was carrying Stephen and we bought our first residence, a 10 x 20-foot garage on a block of land in Chullora, Sydney. They used to call them 'temporary dwellings'.

After the war there was a building shortage all around the country and not many people could afford the total price of a house and land. Many people, like us, purchased a block of land first and built their house bit by bit as they could afford it. This happened gradually, and councils in Sydney gave permission for owner-builders to live on their land in a temporary dwelling providing their house was under construction.

It was very expensive to buy a house in those days. You had to have 50% of the purchase price as a deposit. You couldn't take out a home loan from the local bank. The only way you could get money was through a building society or a terminating building society.

So when I left Nicholson Brothers & Lucas we were quite financially strapped. Pat and I had very little money when I left. I still owned the little Renault motor car. I sold that straight away and bought myself a 1936 DKW because it was cheap

Clive Nicholson nearly broke us. When we first joined him we had a modern car and we'd gone from that to a 1936 car for heaven's sake! So we had very little to thank him for from what he paid us.

People are full of contradictions, and the impression I have given of Clive Nicholson is that of a tight-fisted boss.

This is far from the total picture. Eight years later, when Pat and I had become his competitors, he re-emerged in my life as a remarkably loyal – and generous – friend.

Between Jobs

And so I entered on another period of my life when I embarked on another range of jobs.

I worked for a while selling International trucks. I wasn't very successful at that. So I went back to being a door-to-door salesman.

At the time there was a firm that supplied hand-operated mixmasters called Kan Wundermixers. In those days mixmasters were difficult to buy and if you did manage to track one down, you had to pay a lot of money for it. Then this Kan Wundermixer came on the market with a handle on the top and a pair of beaters, and a big-time radio man called Eric Baume promoted them on Radio 2GB. I got a job selling them door to door and people could buy them for five shillings down and five shillings per week. Although they weren't very dear, it would take them three months to pay the damn things off.

The first five shillings was our commission. It doesn't sound much today but at that time wages were £4-£5 a week, so I got a pound for every four that I sold.

With this radio advertising going on, they shouldn't have been difficult to sell, but I couldn't sell them because they were such disastrous things. You almost had to put your foot on the damned thing to hold it in place, and then as soon as you put it down onto something, it became impossible to stir, because if you stirred too vigorously the whole gadget would start moving around the table. I tried selling them for a few weeks, but I wasn't successful.

By now we were desperate for money. To makes a bad situation worse, when Pat went to St Margaret's Hospital to have her pregnancy check-up, somebody broke into our house in Chullora and stole what little money Pat had put away for our monthly payments. On the same day, while Pat was at the hospital, someone took her handbag and pinched her purse that contained all the rest of her money.

The purse was later found in the toilets, empty except for her house keys and some receipts for some lay-bys for Stephen's baby clothes.

Taxi Driver

I needed another job desperately, so I responded to a newspaper advertisement that said, 'Wanted Taxi Drivers - immediate start.' The depot was at Marrickville, so I jumped in my car, drove to this place and Ken France, the owner, said, "When can you start?" Ken owned three taxis, all with RSL Cabs, and he offered me one to drive.

I said, "Well ... I haven't got a cab license."

He said, "Don't worry about it, I can organise that tomorrow." So he rang up a friend and I got my taxi driver's license on the spot, as I was already a holder of a public vehicles' license.

I started driving Car 42 on Good Friday in 1955. I don't know if it was the 15th day of the month, but it could have been. I turned up at 2.00 in the afternoon and he handed me a little leather bag. Then he showed me how to work the meter because in those days the meters worked by clockwork and we had to wind it up.

I said, "What do I do now?"

"Get out there and earn some money," he said, so I took off.

I was driving along Marrickville Road and somebody waved me down. I pulled up and the passenger said, "Take me to..." and I can't remember what he said, but I didn't know the place so I said, "Where's that?"

He said, "I'll show, you go."

So that became my saying, "*You show, I go.*"

I continued to drive taxis full time for the next 18 months. I used to work six shifts a week and I'd have Mondays off.

For some years I held the record for the greatest amount of money earned over a three-day period. It happened this way: one Sunday morning, I was driving through the city looking for fares and a radio call came in, "Go to the Wentworth Hotel."

I turned up just as the man I was supposed to pick up was getting into another cab.

Before I had time to feel disappointed, someone else walked out of the hotel and hailed me. "I want to hire the car," he said.

I said, "Sure."

He said, "I want to use the car for whatever I want, and I'll pay you one New Zealand pound per hour."

"Whoah!" I exclaimed, "I'll take that on!" (At that time a NZ pound was worth 25/- Australian, and here he was prepared to pay one pound per hour, every hour.)

The reason for his party spirit was that he had sold his business in New Zealand and had come to Australia with his wife to have a wild time. In those days you couldn't go to a hotel on Sundays and buy alcohol, there was none around, so you had to go to the sly grog shops. The first place I took him, was the famous Thommo's Two-Up School where I purchased sly grog. Every cab driver knew Thommo's. It was virtually next door to the Paddington Police Station, and they would sell you as much grog as you liked.

He bought a bottle of whisky for himself and a couple of bottles of beer for his wife. He never used a conventional bottle opener; he used his teeth to take the tops off.

On the Sunday I stayed with him all day. We went to the Taronga Park Zoo, Manly Beach, up to Palm Beach ... I drove him to anywhere he wanted to go.

I had him and his wife for three days and two nights. At one stage he wanted to take a look at Broken Hill, then he changed his mind.

We parted company at Wollongong, and I had lots of cash.

Uttermost To The Guttermost

Sunday morning, I'd go from the uttermost to the guttermost. I'd drive out to the western suburbs where I'd pick up people who I would drive to church for 10/-, then in the afternoons I'd pick up call girls from Kings Cross and drive them to their respective destinations.

A message would come over the radio to say they wanted four cars for Cathedral Street, Forbes Street or Bourke Street, and every driver knew that meant that the girls were on their way out.

Each girl always paid us one pound each, so we knew we were on a £5 job, which was quite good money in those days. (We were clearing about 8/- per pound to drive the cabs and the owner would get the balance for supplying the car and the petrol.) It would take about an hour to drive the five girls, which was why it what good money. Of course, the girls didn't pay the cab fare themselves. The girls just put it on the client's tab.

The only real problem with doing these runs was that they were illegal and the coppers used to threaten us with mayhem because they knew what 'five girls in a car' meant, especially dressed up the way they were on a Sunday afternoon. Even though we were no more than taxi drivers, driving them to a job was classified by police as 'living off the immoral earnings of prostitutes' which was the crime we were supposedly committing.

If the coppers wanted to get heavy-handed, they could do so easily. I nev

er got caught, but some of the other drivers did and we would know all about it when it came over the radio. "Bloody Bill's got caught for 'living off'!" That's what we called it, 'living off'.

It was all part of the game; there was no big deal about it. The same game still goes on today – only worse – and everybody knows exactly what's going on.

If the girls were picked up by the police and taken to Long Bay, they'd keep them overnight. They would pay their fine the next day, or somebody paid it for them. Part of the deal was that when they were released they would have to be driven back home. So sometimes we'd hear over the radio, "Four cars needed for Long Bay". We would drive out there – four to five cars in a row – we'd pick them up and bring them back to Kings Cross. The Police knew exactly what was going on. In fact, it was a comfortable situation and everybody worked together – the coppers, the call girls, the pimps and even we cab drivers.

Part-Time

After a time I stopped driving full time and I went on a part-time basis. I was promoted to running the place for Ken France who had moved to a big service station in Canterbury where he had 30-40 cabs. He paid me the same money, plus holiday pay, as he would have paid if I had been driving a cab. Just working a few pumps was a lot easier than driving.

The cars came in, did their changeovers and they always had a wash, so they got washed twice a day – at 2.30-3.00 in the afternoon and at 2.30-3.00 in the morning.

Ken France's wife also gave birth to a son in the same week as Stephen was born.

Stephen

I was still a part-time driver when Stephen was born on 18 July 1955 at St Margaret's Hospital, which had a very good reputation for maternity cases.

Pat had been in hospital a few weeks before her labour started, and we were very lucky that Stephen was a full-term baby. Pat had been three days in labour and was having a very bad time. The doctors tried to bring the baby on a few times but they had not succeeded.

I'd been continually visiting the hospital for those three days, and when I could not be physically present I was on the phone at every opportunity.

When Stephen was born, I rushed to see him and Pat. I parked the cab in Bourke Street, Waterloo, near Taylor's Square. Ken gave me a cab to drive as transport that day. That's the sort of person he was.

Ken was a very good friend. In fact, in later years it was he who created the opportunity for us to get Stephen into Newington College.

CHAPTER 9

Superior Safety
Salesman

Before I met Pat there had been times when I was penniless, friendless and without close family connections. With Pat I had a home, a son and a constant source of income. How could I have guessed that my marriage partner was also my ideal business partner, counter-balancing my own creative skills?

Looking back I realise that from the time of Stephen's birth, every year was onwards and upwards, and 1957 was no different. We were about to enter the make-or-break period of my life; I was on the way up.

How could I have known that the years I had lived in western New South Wales were spent working alongside the type of people who would someday become my Driza-Bone target market?

How could I have imagined that my time spent working in the safety sector for Nicholson Brothers & Lucas would ultimately lead me into protective oilskin rainwear?

And how could anyone have foreseen that the shearing sheds, horses, sheep, drought, rain, engines, factories and country people gave us an understanding of the Australian consciousness that would prove so useful to Pat and I to make a success of the rest of our lives?

For the rest of my life I always seemed to finish up on top of the heap. Sometimes I even think, "Did somebody guide me? Was there a master plan laid out?" Yet at no stage did I ever sit down and write out that master plan. I never said, "In 30 years from now I'm going to be a multi-millionaire." It never entered my mind. I have been successful in many of the things I've done in my life without knowing how it was all going to happen.

However, I did want to be a success and Pat wanted to be a success too. Perhaps this was the unspoken agreement between us and one of the reasons that made our marriage the happy union that has stood the test of time. We were both prepared to work at success. We were prepared to work the long hours and do whatever successful people do, but we didn't set out thinking, "This program we are now entering into is going to have this sort of result."

However...

Ex Mat Man

As far as I could tell, there was no future in being a cab driver, so in 1956 I decided to look back on my life and re-evaluate it. From farming to printing, from radio manufacture to face-to-face selling, I'd had many experiences. Some things I was good at, other things – not so good. However, I could not get out of my mind the fact that the industrial safety sector was in its infancy in Australia, I had acquired an understanding of many aspects of that industry and I would do well to return to it.

When I picked up a glove, it was clear to me how it was made. I could not only see the stitching across the second and third digit, I also knew why there was a quarter inch of stitchwork where the thumb joined the glove. This understanding came to me. I understood not only the precise measurements inside protective headwear, but the reasons why two inches was a safe distance in the inside casing of a hardhat, and why one-and-a-half inches was unsafe. And I could analyse goggles, steel-capped shoes and protective rainwear in the same way.

Despite the fact that I liked my association with Ken France, I wanted to stop driving taxis and get back to industrial safety. I still knew some people in the safety business, so I contacted them, and they told me a firm called Superior Safety Suppliers Pty Ltd was looking for a salesman.

Superior Safety was owned by John Partridge. He and his wife Sheila worked from their home in Haberfield, Sydney. So I applied for a sales position with him, and I was accepted. Stephen was about nine months old when I joined Superior Safety and I remained with the company from late-1956 to 1961.

John Partridge was a big man – and I mean a *big*. He was very noticeable. He was an ex mat-man – a former wrestler – who weighed in at 20 stone. He was also a brilliant salesman. His advertising campaign was a 'WANTED' pamphlet with a picture of himself captioned with the words, 'Would you buy safety equipment from this man?' That was brilliant, because everybody recognised him at once as someone they would want on their side! But John wasn't just hype, he also knew safety and safety equipment very well.

Snow Mountains and The Sydney Opera House

The Snowy Mountains Scheme was launched by the Government in 1949, with 65,000 people from 40 countries, together with 35,000 Australians, constructing the world's biggest and most complex hydro-electric scheme on schedule and within budget. Superior Safety supplied the first Australian-made safety helmets to the Snowy Mountains Authority. Before we came on the scene they had been using American and English product, but John sold them the first of the fibreglass helmets. He used to travel to the Snowys from Sydney once a month and Superior Safety met most of the Snowy Mountains Authority's safety equipment needs.

And it worked. At that time the worldwide average loss of life in hard rock tunneling was three men per mile. The Snowy average was one life per mile of tunnel. In fact, the six-and-a-quarter mile tunnel from Jindabyne to Island Bend set the world record, which has never been beaten – one mile in 66 days, two miles in 141, three in 216, four in 288, five in 370 and six in 468 days – without the loss of a single life. And Superior Safety played its part in this outstanding result.

In 1954, the Premier of New South Wales, the Hon J J Cahill MLA, announced the appointment of a committee to advise the Government on the building of 'an Opera House'. In 1955, Cahill announced that an international competition would be held, open to architects of every country, and from the submissions received, the New South Wales Government would select a design worthy of the Commonwealth's finest Opera House. In January 1957 the prize was awarded to a 39-year-old Danish architect, Jørn Utzon. And when the preliminary work began in 1958, Superior Safety was on the spot contributing to the Opera House's safety standards. We were down there regularly. John and I sold them much of their safety equipment, particularly their rainwear.

By this time Pat and I had sold our temporary dwelling in Chullora, and we were struggling to buy our first house in River Road, Revesby. It was a brand new eight-square fibro home. We had scrimped and saved until we came up with the required deposit, which was half the price of the house. And then we borrowed the other half from a terminating Building Society.

In 1958, Pat and I moved to Burns Road, Picnic Point. After the eight-square fibro place we had down in Revesby, we thought we'd done ourselves proud. Our new house was about 12-13 squares – and *brick!*

The Picnic Point development was among the first of a style of housing development that is commonplace in Sydney today, where the developer sub-divides a block of land and builds all the houses. The developer was a man named 'Skan' and he also used to run a firm called Rotofridge, which pioneered the ill-fated round fridge. When you opened the door, even the shelves were round. I couldn't help but admire the design although – like most people – we never got one because a round fridge is completely out of place in a rectangular kitchen.

Beyond Cottage Industry

John Partridge was a good salesman who knew his business and he taught me a lot about selling. By this time safety equipment had grown beyond being the cottage industry. But it hadn't grown massively.

As I said, selling safety was almost like selling insurance. People buy insurance because they have to – they don't buy it because they want to. They always say it's too pricey. They get nothing tangible for it, and the best insurance you can have is the one you don't use. The same applies to a fire extinguisher – the best fire extinguisher you can ever buy is the one you don't use. Likewise, the best goggles, respirators or safety boots are the ones that save somebody's eyes, lungs or toes. John understood that and he could put that story over very well.

He didn't manufacture at this time. Superior Safety was a distributor for Nicholson Brothers & Lucas, Protector Safety and Nu-Plac.

John and his wife Sheila lived in a house in Haberfield and they ran the business from their home. During this period I sold my car because John gave me a company car, a Standard Vanguard. He paid all my running costs, which was a vast improvement on my arrangement with Nicholson Brothers & Lucas.

Another factor that made my working life easier was that the unions had negotiated much better wages and conditions for workers. These included the 40-hour week in 1948, long service and sick leave in 1951, and continuous improvements in the mid-1950s, including pension schemes, superannuation and many other policies which took effect at different times in their respective States.

Television

Another innovation was the introduction of Australia's first regular television service in 1956, black and white of course. This had an enormous impact on the nation's social habits, as well as on all other forms of entertainment, including radio, dancing and music. For the first time, the family lounge area was becoming what is now known as a 'home entertainment centre'. However, Pat and I did not race out and purchase a TV. In those days everybody wanted one, so there was a long waiting list. People had to wait up to six months to buy a TV.

The reason we got our first TV when we did was because I did a good turn to someone. It happened this way: immediately outside our front footpath in Revesby was a dirt gutter, and one night a man bogged his car in it. I could hear the commotion outside my front gate so I went out, helped him push his car out and I got splattered with mud. The next day, when this man came back to thank me, he told us that he was the local TV retailer and he offered us a brand new TV straight away if we wanted one. Furthermore, he said that he could use his influence to organise our repayments through the local bank at an advantaged rate, and we accepted his generous offer.

Sometimes you do a good deed without ever imagining the significant affect it will have on your life. We became very popular in our street because we had a black and white TV before they were commonplace.

And so, instead of listening to *Dad and Dave* on the radio we got to watch all the latest shows: *Death of a Wombat*, *ABC Weekly*, and – although we didn't like it much – Brian Henderson's *Bandstand*. I remember seeing Johnny O'Keefe singing on the show. Although he was a bit way out for us, his music was interesting at the time, because we felt you could sing along with his songs.

During this period of our lives Pat and I would go to dances irregularly; however, we would play tennis one night a week. But most of the time we were busy going to work.

I still played my banjo-mandolin on the odd occasion, especially when we went out – for example, I might take it along to a tennis night. Although I sang badly, Pat is a good singer and every now and then we would get together and sing songs with a group of friends. (Pat could also play violin but she gave it up soon after we met.)

Protector Safety

John Partridge eventually moved the whole Superior Safety operation from Haberfield to Bankstown. John and Sheila bought a big house on a big block of land in Chapel Road, which today is right near the heart of Banktown's business and shopping district. We helped them set up an office in their double garage; we also used part of the house as a storeroom and we took off!

One of our most important business alliances was with a company called Protector Safety, called 'Protector' because their product protected people's eyes. Protector Safety Pty Ltd was owned by a man called Herman Dotch, whose name originally was Deutsch. He was a German Jew who came to Australia in 1937 to escape the Nazis. (His son is quite well known in financial circles even to this day.)

In 1938, Herman Dotch started importing sunglasses from the continent and he expanded into goggles, lenses and optical wear. During the war years his business had been taken over by the Government to assist the 'war effort'. When we came in contact with him, his business was in Sussex Street, Sydney. He had a singled-fronted, three-storey building – now long gone. It was decrepit but he had acquired it during the war years and had remained there throughout the 1950s.

He used to import stacks and stacks of sunglasses and he would sell them all over Sydney. He paid very little for them and sold them to optometrists, who would mark them up by 100 per cent, so it was an extremely profitable business. Although Herman's main business was importing and distributing sunglasses, by the time I started with Superior Safety, he was also manufacturing goggles and spectacle frames.

Herman and I became friends. Over the next few years I spent a fair bit of time working with him introducing various types of safety equipment into the market. Herman relied on Superior Safety to be the salesmen on the road. I knew what the job required and I would come to him and say, "Herman, this is this company's problem, what can we do about it?" and he would think of appropriate solutions to

Mum and Dad (Gilbert and Claire Fisher) before I was born.

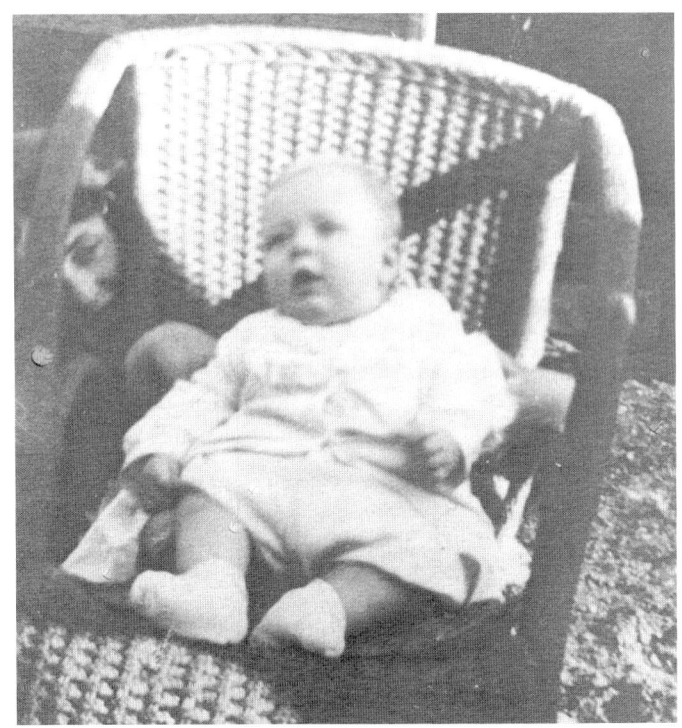

Me as a baby, in 1926.

Waterford Station, Peak Hill, (L-R: Tommy Rice, Shirley and me) in front of the 1928 Chevrolet.

Dave, Mum, Shirley and Janet in 1943, at Circular Quay, going to Manly for a Picnic.

Shirley and me on Carbine at Peak Hill.

Pat and me on our first date.

Me.

Pat.

Signing the marriage register.

Stephen and me at our Driza-Bone factory in Guildford.

Pat and me in our Driza-Bone coats.

Stephen, Pat and me inspecting a 50-year old Driza-Bone, returned for reproofing.

Signing away the company. Pat, me and Stephen Knight, financial director of the Halstead Group.

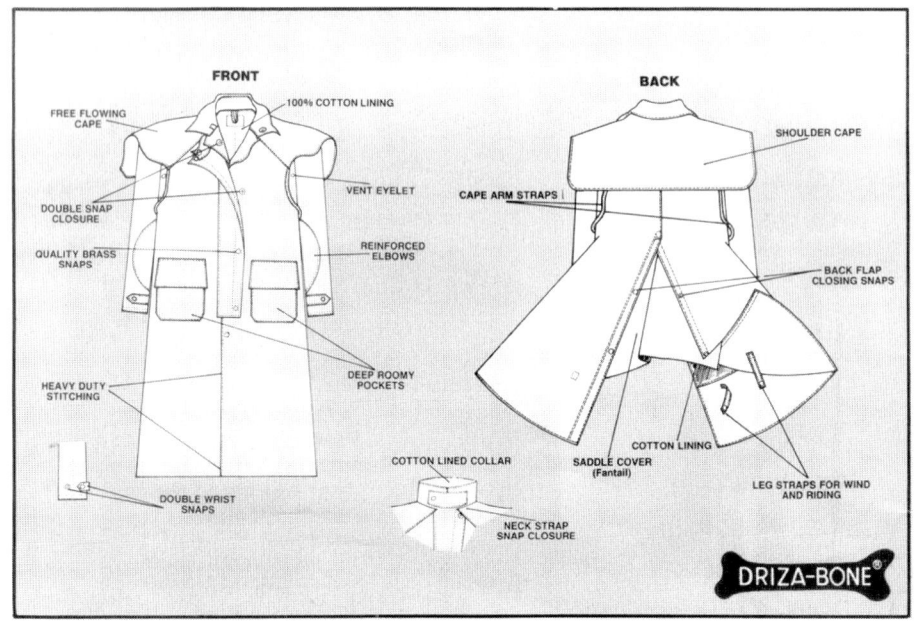

What comprises a Driza-Bone.

Driza-Bone promotional material.

make a product to handle the job. I'd then take it back to John for his approval and Superior Safety would sell it to industry.

For example, an industrial manufacturer might want a pair of goggles that protects a welder's eyes. This is not a single solution because one pair of goggles is needed for grinding, another for dust and yet another for welding, so that's three different types of goggles. Then there's goggles to fit people with glasses – that's six goggles. Then there's goggles for steel manufacturers, goggles for spray paint-ers, goggles for construction workers, goggles for ... oh, the list goes on and on! So it doesn't take long before you're talking about almost a hundred different types of goggles. And the same thing applied to respirators and to headwear. Then there is 'hard' safety equipment, like hard hats, and 'soft' safety equipment like rainwear and gloves, which added to the complexity of every requirement.

Because of Cliffy Gartrell's contribution to the setting of new safety standards, John and I convinced Herman to make the first fibreglass safety helmets. This was a real innovation because not everybody knew how to handle fibreglass in those days. It was a new invention, requiring a specialised type of manufacturing operation.

Herman had to design them, then make the frames and the moulds. Furthermore, each one had to be made by hand because the liquid fibreglass had to be poured into it, in different colours and different sizes to fit the specific requirements of each job. So quite a lot of labour content went into making each hat.

Although I understood the design and manufacturing process, I never got involved in the actual production. Selling these new innovations to the companies who needed to use them was my job. I was at the sales end of the marketplace, which I found to be a most satisfying role. I was also gathering ideas from the respective factories and construction businesses to enable us to create a product that would match their individual requirements. I am proud to report that we did it right. In fact it got to the stage where John Partridge and Superior Safety Supplies became the first A-class distributor for Protector Safety.

Protector Safety P/L had never set up three levels of distribution before, so the A-class distribution returned a very significant increase in profits, allowing us to sell to distributors and resellers, while still making a good percentage ourselves:

- An A-class distributor got $33^1/3$% discount
- A standard class distributor got 20% discount, and
- A reseller class distributor got 10% discount

This gave us a $13^1/3$% profit on standard sales and $23^1/3$% on reseller sales, therefore – for the first time – we did not have to actually sell the goods ourselves to make a significant income.

Money Management

As a businessman, John Partridge had one big fault. He didn't put any value on money. He was a very good salesman but he could never understand how to balance his books. If he had money he would simply spend it. Intelligent money managers understand that there are different types of money. There is personal money, turnover, investment, working capital and so on. Well, all this was beyond John's comprehension. He couldn't tell the difference between personal money and business money, and an awful lot of small businesses have that problem today.

Instead of controlling his money to make the business more substantial, when John would see that his bank balance was in the black, he would simply go out and buy another car. So he had lots of cars. Through his mistakes, I learned a key business lesson about money management, and when we started our own business it was

Pat who implemented this lesson in a practical way – because she already had this understanding. But thanks to John, when the time came, I understood the necessity of Pat's monetary policies and I didn't fight them.

However, these cracks in John's business were not apparent at this point of time. We had earned our A grade rating, everything appeared to be going extremely well, and I was learning more and more about how to be a good salesman. Business was booming and the sales team expanded. In 1961, John had to put on another salesman – Digger Brooks – who didn't have a lot of experience in selling, but quickly learned the ropes and became good at his job in his own way. Next we added an American fellow while I was promoted to 'Sales Manager' of a sales team of two.

Despite John's title of Managing Director and mine as Sales Manager, we didn't really have formal positions – we had a job to do and we just did it. And that suited us both. The only thing that was beginning to aggravate me was that John was always short of money and he was always unnecessarily late in paying his accounts. After having achieved so much over the five years I had been working for him, Superior Safety was never more than a borderline business.

Over the years – with the assistance of Cliffy Gartrell, Herman Dotch, government contacts, the various regulating bodies of which I was a foundation member and many of my colleagues who had played a role – we had moved to the forefront of our industry. Personally, we all earned a comfortable income – but why the hell was Superior Safety more or less still borderline? I couldn't understand it.

Gilbert Fisher (1903-1961)

My father died in September 1961. I had booked an appointment with a customer when I received a call from the hospital to say that my father had been checked in. However, they failed to tell me how crook he was, and so, regrettably, I fulfilled my appointment before going to see him. When I arrived at the hospital, they told me it was too late. He had already died.

My father died broke, and for a while it looked as if Shirley and I would have to bury him in a pauper's grave.

Shirley and I could never trust our father. Shortly before he died he told us that he would be leaving us a considerable sum of money. He also told me that he owned the largest building in North Sydney – the Greater Sydney Building Company Pty Ltd, located across the road from the Post Office – but, of course, he didn't.

I have always had mixed feelings about my father. He had many advantages in his early life, which I never had. Until the tragic death of his parents, he had obviously enjoyed a comfortable upbringing and a good education. He had a certain style and I resent his inability to capitalise on the natural advantages he was given in life. My father wasn't a hobo as such, but he certainly wasn't a successful person.

He wasted his life.

Of course, Shirley and I were greatly saddened by his death.

On one hand, we were sad because he was our father.

On the other, we both mourned what 'might have been'.

Lungrens Gloves

John wanted to add gloves to our product range because he realised that even as an A-class distributor, we weren't making the right amount of money based on the investments that we had. (Looking back, another reason he didn't do as well as he should have, was that he never had enough capital on hand to get the best buying advantages.)

I see this with hindsight, but Pat could see it at the time, because John had asked Pat to come and work with us a few times, so she got to know the business from the figures. It wasn't like having a stranger enter the business; we had been working for some years and our families knew each other. We had gone boating together and done all sorts of things together as families. Pat and the accountant got along like a house on fire, but John would never take any notice of his accountant's advice or of Pat's. Doggedly, John simply couldn't – or wouldn't – come to terms with any of the figures.

Whenever Superior Safety got into trouble John would ask Pat to come over and make the figures add up.

So Pat would come over and try to convince him to do things the right way. Pat accounted for everything; she wouldn't even leave 5c out. She tried to introduce that principle to John Partridge, who shrugged her off with the words "If it's close enough it's good enough. Hell's bells! Don't worry about it!"

By this time the dire consequences of his refusal to manage his money were almost inevitable. The lesson that I learned from working with John was to never take a gung-ho approach to money management that I have never done in our whole time in business.

At the time when Merrylands was a fast-growing suburb of Sydney, there was a firm located there called Lungrens Gloves. Don Lungren made industrial safety leather gloves and in 1961 he wanted to sell out.

So he looked up John Partridge and told him so, and John said, "Yes, I'd like to buy the Lungrens Glove company."

So in 1961 John did what many businesspeople do when they sense they are going to the wall – he bought another business to prop up the first. Superior Safety bought Lungren Gloves and John convinced two silent partners to put up the investment capital.

Shortly after this John came up to me and said, "Frank, I want you to take over the glove operation for us. Spend some time at the Merrylands plant and learn how to run the business over the next three months before Don Lungren leaves."

I said, "okay".

It seemed like a win-win situation. I had just become financially confident again. Pat and I had decided to buy our own car and give the company car back to John. John had agreed to pay me a car allowance that, in turn, paid for the running of the car and actually paid us a small profit. It was a good arrangement for John as well, because it freed up the money he had previously invested in the car, which gave him more ready cash – to *spend*.

Here's another lesson that became apparent to me: instead of locking capital up in motor cars, it's better to have money as stock. After all, Superior Safety Supplies was making profit on its stock, not on its company cars.

So John sent me over to Lungren Gloves where I started as their dogsbody. It was the first time I had ever seen a sewing machine in pieces, and my mind started looking at this forlorn aging equipment, wondering whether or not the place was running efficiently.

I applied myself and quickly learned all about sewing machines because Don Lungren was a good teacher. He taught me that the right type of sewing machine to buy was the Singer 3IK 20. These are post-War industrial sewing machines. They were the best of the type for making gloves then and they still are today. Singer stopped making them just after the war. The last ones were made in 1953-54.

So I started working at the Merrylands plant and learning from Don until Christmas came and we started closing down for the holidays.

I didn't know that, in the interim, Superior Safety had gone broke.

I'm Sacked!

As we wrapped up the year's work and prepared to close for the Christmas vacation, John asked me to step into my office for a word in private.

He sat me down and sadly said, "Frank, we are dispensing with your services." He handed me a pay envelope with those parting words.

I got the surprise of my life when John told me that Superior Safety was in liquidation and declared bankrupt for £26,000. (In 1961, £26,000 was probably worth $1 million in today's money.)

So for Christmas 1961, as my entitlement I was handed two weeks' wages instead of the two weeks' holiday pay.

I wasn't the only one who copped it either, because when John went out of business his two silent backers suddenly had Lungrens Gloves dumped back on them. They were two old fellas who didn't know the difference between a sewing machine and a typewriter, for heaven's sakes!

They suddenly found themselves having to buy out John's share of the business, because under the partnership arrangement if one wanted to sell out, the other two had to buy him out. So these poor old investors had to buy John's third share in the partnership, and their money went straight to the liquidators.

CHAPTER 10

Our Own Business

So for Christmas 1961, I copped a sacking, by which time I was so well known in safety that a decade of my life would be wasted if I left the industry.

There was clearly only one sensible move Pat and I should make, which was to start our own business. This opportunity was staring both of us in the face, and even though we had discussed this idea in a very vague way on and off over the years, we still needed someone else to make the suggestion.

Harry Guest and Patricia

Pat had previously got herself a position working as an accountant for a firm of bed base manufacturers called Usher & Guest Pty Ltd which was owned by Ray Usher and Harry Guest. In time, we got to know Harry quite well.

Harry was just a nice ordinary man. He wasn't tall, but he wasn't short. He had reddish, curly hair, and nothing else about him really stood out. Ray Usher and he had served in the Army together which is where they decided to get together after the war and start a business as partners.

Pat had only been at Usher & Guest for about six months when I got sacked. Harry spotted the sad look in Pat's eyes and asked her what was making her so unhappy. She then walked him through the whole Superior Safety mismanagement saga, culminating with the fact that my Christmas present was a sacking. On top of this, she explained, we were stuck with having to find the money to pay for the brand new Holden car which we had only just ordered and which would no longer be subsidised.

She told him this tale of woe and after listening attentively and asking a few questions he said, "Why don't you start your own business? With Frank's contacts, it's certainly possible."

With 10 years' experience in safety behind me, it certainly seemed like a reasonable suggestion. I had the contacts, the knowledge and a good understanding of how the sector worked. However, I had little time to lose because my two weeks' retrenchment pay was fast running out. So instead of taking a vacation, Pat and I got our heads down and figured out how to start our new business.

Harry was so confident of our chance of success that he provided the start-up capital to kick-start our business. He gave us a few hundred pounds that, at the time, was sufficient to get us moving. (He didn't actually have the cash, so he raised a loan on his insurance policies to give us our start.) Harry gave us all his support and Ray Usher didn't get involved and didn't object. We had very little to do with Ray; he just went along with Harry's decisions as far as we were concerned.

Isn't that amazing? Harry offered £600 so that we could start a business with him as a 50% silent partner. For our part, we put up the car and my time, as well as my father's old printing press that I had inherited. And so I could do my own printing from home. Harry must have foreseen that once our business got profitable, he would obviously be losing a good employee in Pat. Pat was managing his books better than they had ever been managed before.

Financially, book wise and customer-wise, Pat had become a most important part of Usher & Guest Pty Ltd. She had straightened his business out, and for the first time they too were running at a profit and keeping the partners happy. Yet Harry put aside his self-interest, and rather than slow down Pat's career path, he chose to give us our start. It was very decent of him. He even allowed Pat to use his company telephone to take our orders over the phone. So our first business phone number was Usher & Guest's.

To be honest, Harry had a lot more confidence in Pat than he had in me. In fact, he didn't know me very well at all, even though at the time I had a fairly high reputation in the safety industry and Harry knew that. I had been in it longer than most people had. Harry knew that too.

However, Harry had some inside knowledge of my expertise because he used industrial safety equipment in his business, including welding equipment and other metalwork-related gear, like wire weavers. He had quite a collection of dangerous-looking machines on his factory floor, so he had some industrial safety knowledge.

Harry was always a silent partner in our business. We used to hold formal board meetings every so often, and we'd also contact each other quite regularly during each week.

Our Name

The 15th of (that 15 again) January 1962 was the official start-up date for Armour Safety Pty Ltd. We started with Harry's few hundred pounds and his confidence behind us.

And so, we had to call the business something. We sought two qualities in a name:

1. We wanted something as close to the beginning of the alphabet as possible (in the 'A's or 'B's). This had nothing to do with getting an upfront listing in the Yellow Pages. It was because when the end of the month comes around and the accounts get settled, the debtors who owe – say – $4000 but only have $500 to pay it with, will pay the ones in the front of the queue. Therefore A gets paid before Z.

2. The second thing we wanted was a name that was not our family name. We didn't want to call the business Frank Fisher Pty Ltd or Pat Fisher Pty Ltd. Instead, we wanted a name that was descriptive of the business we were in because we wanted our products to stand on their own two feet without being tied to the founders, in case we ever wanted to sell out.

Furthermore, we didn't want to use this trick of using 'Aardvark' and all those sorts of stupid words. We simply wanted something that was descriptive of our business that preferably began with the letter 'A'.

So Pat got out her dictionary and she came across the word 'Armour'. This particular dictionary said, "Armour is any protective clothing or device worn by a person whilst engaged in their trade or profession." So what was a person who wore a suit of armour? He was a professional – that's what he was. Why did he wear armour? So he would be protected. We were selling industrial safety equipment; what did we call ourselves? Armour Safety Pty Ltd. It was a description of what we sold. Think about that. What we were selling was exactly the same sort of thing – in other words, we would be selling 'industrial armour'. It had a certain marketing ring to it –as we found at our first trade fair.

It didn't matter who was doing the selling – whether it was me or someone else, there were no personal name ties – the product stood on its own merits. And sales were always more important to us than putting the Fisher name forward.

The word 'armour' evokes knights and jousts – which is glamorous; the words 'industrial safety equipment' conjure up the image of industry, which is dull. The more we promoted the idea that we were selling industrial 'armour', the more curious people became. People wanted to know, "Where do I get my industrial armour from? Who was that bloke who was telling me about it? Oh – Frank Fisher, we'll get it from him and Armour Safety." It was very effective. It's a beaut name.

And that's how it was with Driza-Bone. We could have renamed it, *Pat, Stephen and Frank's Driza-Bone*, for heaven's sake! But we insisted it be just *Driza-Bone*. We stepped back and allowed the company to have its own reputation without a personality image attached to it. Driza-Bone was its own product. When the time came to sell the business, it would not have been as saleable otherwise.

It was a mistake R M Williams made. R M Williams used his name and Akubra doesn't (ie. it's not the 'Keir' hat, which is the family name, its name is the Aboriginal word for 'head covering').

R M Williams sold out before we did, and he was paid $14 million for his business. People said at the time that the company was undervalued by $10 million because the company name was intimately tied up with R M Williams, the person. (They're doing much better now that they have changed their emphasis from R M Williams to 'RM's', even though everybody knows what RM stands for.)

Starting Out In Business

By this time Superior Safety Supplies had gone into liquidation, so I took over where I left off in my previous job. I went to the same customers and I didn't have to re-introduce myself. I got our first order on day one.

Thanks to Harry permitting Pat to accept Armour Safety phone calls while at work, she made a big contribution to our start-up and she would put in a lot of extra hours to make up for the time she spent working for Armour Safety on Usher & Guest's premises.

I found it quite amazing that we made a small profit at the end of our first three months. I've kept that balance sheet. We made a profit because I didn't take any wages; all I took was £10 expenses per week. We started off the same way every small business does: you don't pay yourself if you don't have the money. After three months I started paying myself a wage of £10 a week.

This arrangement also solved our problem of having taken delivery of the Holden. The car became an important part of our business, working as a delivery vehicle as well as sales transport. Furthermore, the car was being paid for by Armour Safety Pty Ltd.

Company Colours

In the early-1960s General Motors Holden released a few cars in two colours, and we had bought a red one with a white top, so red and white became our company colours. It was quite a distinctive car. My very first business cards were red with white printing on them.

My first order was a firm called Rocla Pipes Pty Ltd in Belfield who bought some gloves. My second customer was Structural Steel Pty Ltd, which made the reinforcing steel mesh that goes in concrete.

My next customer was the British Leyland Corporation when it was known as the British Motor Corporation (BMC) or Morris Motors. In those days the cars came out from England in pieces and were assembled at their new plant in Zetland. I had previously supplied them with some safety equipment when I was working for Superior Safety Supplies, so I continued our relationship.

The British Motor Corporation was a significant account. Shortly before its demise, Superior Safety Supplies Pty Ltd had been offered a contract to supply industrial safety equipment to the new CKD (completely knocked down) plant built on the old Rosebery race course. I made my first call on the company when the main office was in the old jockeys' building and it still had the aroma of its past history. I started supplying equipment to British Leyland, particularly safety footwear, gloves, aprons, respirators and goggles.

Over the next few years I spent considerable time offering them advice on safety procedures and I became one of the few outsiders who had access to all parts of the plant, including its secret developments. A number of purchasing officers were appointed to look after safety but none had a comprehensive knowledge of safety until Archibald (Arch) Wicks joined BMC having left Australia's only aircraft company as coordinator. He brought with him an extensive knowledge of safety. He and I found our combined experiences a great asset in running the safety program of the motor company.

After some years Arch retired and left behind a legacy of procedures that continued until British Leyland decided to stop manufacturing in Australia. Over the years from 1962 to 1989 I participated in many similar programs with a great number of companies of all sizes and complexities within Australia. To have been part of that history is a story on its own but suffice to say I obtained great satisfaction from my contribution to safety in industry and the saving of injury to workers and even in saving some lives.

Your Man Of Armour

From the start, Armour Safety attracted good publicity. In the early-1960s trade shows were held at the original Sydney Showground near Moore Park. Shortly after we started in business, an 'Engineering Exhibition' was held there and we wanted to be a part of it. The problem was that we didn't have any money for promotions. We could afford to rent the space, but we couldn't afford the furnishings and props, which most companies would rent from the exhibition promoters. We couldn't even afford a display stand.

So Harry Guest made me the frame, which was relatively easy for him to do from his bed base manufacturing facility. He made it out of 16 pieces of tube about seven feet tall and seven feet wide. The space of the actual exhibition stand was 20' x 10' and I bought some striped red and white curtain material to promote our company colours. Next, I went to an auction of a shoe factory in Hurstville where I bought 100 boxes made out of plywood which I used for shelves – they were a couple of feet long by a foot wide and a foot high, and I covered them in red material too.

We then put on display examples of all the different lines that we had on offer. By this time we had become distributors for Protector Safety, Nicholson Brothers & Lucas and quite a few other industrial safety equipment suppliers. We were selling other people's gloves, plus we had started making our own, so we had a reasonable sort of a display.

And then I did a really smart thing: I hired a real suit of armour from J C Williamson's. I stood it in the middle of the booth under the words '*Early* Safety Equipment'.

On the right, I displayed a mannequin dressed in ordinary industrial safety – a pair of overalls, safety boots, a pair of goggles, a hat, a respirator, etc, under the words, '*Modern* Safety Equipment'.

On the left, I had a table with a sign above it, 'Your Man of Armour', and that's where I stood and talked to prospects.

So with me standing to the left under that sign, a mannequin dressed up in safety equipment on the right and a suit of armour in the middle – guess what the TV people liked?

"There's Fisher, and he's the man of Armour!" they said. This was gimmicky, and we got our name splashed across the TV coverage of the expo. Our opposition, like Nicholson Brothers & Lucas, didn't get on TV at all. Phew, didn't that come down like a housebrick on our opposition!

They were very disappointed that I got all the media attention, especially because James North – Nicholson Brother & Lucas by this time had become James North Pty Ltd – had also been talking about using the name 'Armour' too at some stage. At the time, I didn't know that they had been toying with that name, so we were fortunate to have got in before them. We had it registered. It was ours; nobody else could use it.

Moving Into Manufacturing

We entered the market in 1962 purely as a distributor and we began manufacturing within a year. It didn't really cost a lot of money because we entered the market in a cost-effective way.

At first, we continued our relationship with Lungrens, having them manufacture the gloves that we sold. Then I set up my first factory, which was about the size of a four-car garage, in Ilma Street, Bankstown.

Next, I sought out Ernie Jeffcote who had a business buying and selling secondhand sewing machines and other equipment of that type. He had roomfuls of sewing machines, plus he had clicking presses, and all the things I required to set up my own manufacturing plant. From Ernie I bought five 31K-20 Singer sewing machines and one clicking press – a machine used to cut out pieces of leather.

When making leather gloves you've got to cut the pieces one at a time, which is one reason why gloves are so expensive in comparison to most other things. A good clicking press operator can cut six dozen gloves per hour – so when you consider each dozen pair of gloves has 144 pieces of leather in them, that's a lot of pieces per hour.

One of the ex-Lungren Gloves girls worked for me part time. Her name was Mary. She would come over whenever we needed her to work. I would cut the leather myself on the clicking press at night and she would sew up about six dozen pairs of gloves a day. Then I added more machinists and our Ilma Street plant finished up with four machinists, and one full-time clicking press operator. After this we decided to move to bigger premises in Smithfield, in Sydney's west.

We bought Harry Guest out some two years later and he was very happy because Pat and I had turned his £600 (or $1200) investment into $20,000. If Pat hadn't been working for Usher & Guest Pty Ltd we would not have met Harry, and we would not have been able to achieve what we achieved.

Frank's Workshop

I did the hiring and the firing of the factory staff, the inventing and the research and development (R&D) in the factory, as well as managing the sales. I was constantly working on new designs on the bench at the back of my office. 'Frank's workshop' wasn't a place set aside for the boss, which nobody else could go into; it was just another part of the factory.

Even so, I had to understand what was going on in the office whether I liked it or not. I was never a great office person. I was 'the factory', and when Pat joined us, she was 'the office'.

Pat became the greatest asset as far as the business was concerned because she was so good at all the financial and office management.

I was constantly thinking up and designing new ideas to try to make sure I could make what my customers were asking for. My job was to create the want and satisfy it, and I guess that's what any small business runs on.

I was fairly well known in the trade; I had been in it for a long time. I had a lot to do with changing or improving the products over the years. I had done a lot of surveys of factories and other workplaces. I was trained as a safety officer. I could talk to safety officers at their level. I could talk to managing directors at their level, because I was a managing director also. *I was it.*

This put a lot of pressure on me, but it also gave me the opportunity to innovate things at the time. It also meant that I had a great advantage over companies who were employing salesmen with none or little experience of safety equipment and/or safety in general. As the boss, I could make on-the-spot decisions when talking to customers.

Heart Attack

There were no telephones in the days when we decided to move from our little factory in Bankstown to new premises at Smithfield. I ran myself into the ground running the Bankstown and Smithfield factories, which is why I had my first heart attack.

Stress was part of the cause and smoking didn't help. Although I was never a heavy smoker, I would smoke 10-15 cigarettes a day –

two ounces of tobacco would last me a week. But I did smoke, everybody smoked, so I just did what everybody else did. It added to the stress on my health and I ended up half killing myself.

I didn't have a telephone when we first set up our Smithfield factory, so I had to keep running to the Post Office to ring Pat, who was still working for Usher & Guest. I'd have to ring her every few hours to check on any orders that had been phoned through to her. We were doing everything ourselves, not only the machining but the paperwork too. We had to do the wages and all those sorts of things, because I had grown to employing 18-20 people. Looking after 18-20 people keeps you busy enough, without having the other jobs too.

And I just ran myself into the ground.

I thought I had the flu. I had been home for a week. I had been sick all day and I was home in bed when I had the heart attack. It happened in the early evening – fortunately for me, after Pat had come home from work.

I felt a burning sensation in my chest as if someone had tied a rope around it, which was suddenly tightened up. I was flushing hot and cold. I couldn't breathe; I couldn't get enough air. I felt terrible. I was conscious and still talking while the pain got worse and worse. By that time I got to the stage where I didn't really care what was happening, everything was such a blur of events. Then, I don't remember what happened next. Everything went blank.

Pat didn't know that I was having a heart attack but she could see I wasn't coping. It looked dangerous at the time. She rang the doctor and made arrangements for me to go straight to Bankstown Hospital. Next, she went to see George and Betty Gell next door to ask them to look after little Stephen. Betty said she would take care of him and George – who was also our solicitor – said he would accompany Pat to the hospital.

I arrived at Bankstown Hospital in the ambulance, I think Pat travelled with me while George drove in his car and carried his briefcase. He arranged to see me in Intensive Care because my Will had to be finalised. George said I was looking well and that there was nothing to worry about and I'd be fine – but he also added, "Please sign here". So I signed the Will with the matron and the head sister in the Intensive Care ward as witnesses.

Being that close to death, I found that I was not frightened of it any more and I've never feared death since.

I had another heart attack five years later when I was 45 and they put me into Hornsby Hospital. Again in 1991 I had a quadruple by-pass. So over the years my heart has not been terrific, but it's still going. I saw my doctor recently and he said everything is working fine.

Generosity Of Spirit

At the time of my first heart attack, Clive Nicholson of the re-named James North Pty Ltd, was in France, receiving an award for his contributions to industrial safety over the years. Clive had indeed made a significant contribution to safety and it was a great honour for him to gain this international recognition.

While I was lying in hospital, somebody must have told him about my plight because he rang his office from France and instructed Ron Millar to assist Pat in any way that she might require, including staff help and in the repair of the sewing machines and the clicking presses. Pat did, in fact, have problems with one of the clicking presses and she did need help to fix it. We were in opposition with him by this time. We were *competing* with him, yet he did that for us!

Despite all his meanness with money when I was his employee, Clive Nicholson was a loyal friend to me, and when Pat came into my hospital room and told me, this news lifted my spirits.

While lying in bed, I had been worrying like mad because I was the only mechanic who could fix the sewing machines and the presses. It was essential to keep the machines going. We didn't have many spare sewing machines and the ones on the shop floor were all working flat out because we had built up a strong little business.

Clive did that for me and Pat and I have never forgotten it. I learned a lot from the generosity of spirit he expressed in offering his support to us in our hour of need.

CHAPTER 11

Industrial Safety Equipment

From our Smithfield premises, Armour Safety grew quite dramatically. It was a brand new fibro and concrete brick building, about 3500 square feet in size, on a corner of Robert and Little Street. Smithfield was a new industrial suburb just being developed at the time. It was so new that when we first went there we used to see rabbits in the main street, which is difficult to imagine today. There was also a creek on Little Street, and on the other side of that creek, there was a dairy farm.

When we moved over to Smithfield there were no telephone lines, septic tanks or sewerage in that area, so it was pretty basic. Today, it looks like another place; the semi-rural environment is long gone.

We also chose Smithfield because businesses in the inner-suburbs like Mascot, Botany, Matraville, Alexandria, Waterloo, were dying. People were moving and industry had to follow. That is what we did, because we were labour-intensive. The workers moved out to the areas where houses were much cheaper. People got a lot more for their money buying a house in Smithfield than in Alexandria. They were bigger houses, made of fibro and with big backyards.

Another reason we favoured this location was that it was right opposite a tannery, which supplied us with leather. We found it convenient to buy leather a bundle at the time from across the road instead of having a lot of money tied up in inventory. The only reason we didn't give all our leather business to the tannery was because he couldn't supply 100 per cent of our needs.

Matters Of Money

In our early days in Smithfield I was the salesman as well as everything else around the place – from machine repairer to staff controller. By this time Pat had left Usher & Guest Pty Ltd to join me in working for Armour Safety Pty Ltd. She too went out, knocked on doors and did some selling as the need arose, and did so very successfully.

To alleviate our heavy workload we eventually employed a factory manager, Bob Woods, as well as a salesman, Kevin Croxen, who stayed with us for some years.

We didn't break any records in our first year or two, but we did make a profit and we have continued to make a profit every year in business. From the day we started to the day we sold out, we have never made a loss in any financial year. In 1962 we made a modest profit, in 1963 we made an even less modest profit, but in 1964 we made a very good profit and – averaged over three years – we didn't do too badly at all.

Laminated Aprons

Pat and I – and all our staff – were new, keen and innovative. We introduced a lot of new ideas and Armour Safety kept on growing. For example, we created an apron made of laminated plastic.

It doesn't sound much, but the average plastic apron used in the plating industry was made of cloth with a plastic spread over it. We introduced a laminated apron that had no cloth in it whatsoever, which meant it didn't fall to pieces when it came into contact with chemicals. That was unique and it came about because a firm in Marrickville had acquired a lot of secondhand plastic material and they didn't know what they were going to do with it. They were fabric spreaders by trade, which meant they spread plastic onto cloth. In a series of experiments, they spread the plastic one way and then they spread it on the other way and in doing so they laminated the entire surface. The laminate ultimately meant that they could do away with the fabric inside it. This made a cheaper product because they could use secondhand plastics and not have the cost of the cloth.

However, they didn't understand the significance of what they had made and when they told me about it I said, "I could probably make aprons out of that", which is exactly what we did. That little innovation became a very big part of the apron business in Australia because it was stronger than cloth spread fabric of the same weight, which meant that it lasted longer.

Hessian Welding Curtains

In a factory that did electric welding, welders had to put a flash curtain around themselves and their job, to protect the eyes of other staff members. The curtain frame would have wheels for ease of movement.

For this task, canvas did a very good job, but it had some significant disadvantages, not the least being that it was quite expensive. As an added disadvantage, canvas doesn't have a long life in the sunlight and it certainly doesn't endure when splashed with acid.

One day I was talking to the manager of a PVC coating firm in Ryde about using some wider fabric and the only thing he had of any width was hessian. I said, "Can you spread PVC over it?"

He said, "Sure." And he did.

He applied it to both sides of the six-foot-wide hessian, which was wider than most plastics of the time.

I asked, "Can you make it fireproof?" and he said yes. So he made it fireproof.

I then asked, "Can you also make it ultra-violet ray proof?" and he said yes. So he made it ultra violet ray proof. And we started making welding curtains out of it.

In this way, we introduced the first PVC-coated hessian welding curtains at half the price of canvas. Welders could leave our flash curtains outdoors for months at a time and they wouldn't deteriorate.

World War II Relics

Over the years many eye protectors have been designed and manufactured but most unfortunately used window glass lens, which was easily broken. Two types of safety lens were made, one from toughened glass like the glass used in motor car windows and one from laminating two pieces of glass, so there was a need for a genuine safety lens.

People recognised that a safeguard against dust and chemicals was needed and they attempted to convert ex-Army gas goggles to offer some protection.

The Government requisitioned millions and millions of these gas goggles and after the war they sold them to Army Disposal Shops, from which they became the most common goggle used for grinding.

However, they had no strength in them. They were made out of very thin plastic. They just fitted over the eyes simply as protection from gas, yet they became commonly used in factories. In fact, there was a whole lot of things left over from the war years that were converted into protective equipment, and Nicholson Brothers & Lucas, Protector Safety, Superior Safety and now Armour Safety, all made a concerted effort to weed out the outdated relics and create state-of-the-art safety equipment.

The respirators that were used after World War II were terrible. They should never have been used industrially. Apart from the fact that they were too damned heavy, they were not designed to keep out the chemicals that factories were using; they were designed to keep out problems related to gas. So they were totally inappropriate, yet people were still wearing these respirators – or 'gas masks' as they called them – and they thought they were doing a good job, even though they were inadequate.

Safety was obviously a very young industry at that time, and again – to replace these relics from the war – we sourced goggles, respirators and other safety equipment from overseas to bring industry up-to-date. The impact of all our work was being felt by the 1960s.

Good For Everyone

I continued to go to factories and give lectures on safety to staff. I got a lot of my ideas from the fact that I spent so many years on farms, and a farmer has to be adaptable, just like a person selling industrial safety. For every new function, I had to develop something to suit. I lived and dreamed it, day and night. And every day I learned something new.

These were exciting times for the burgeoning safety sector. We were all trying to sell safety equipment. The safety equipment paid our wages. Selling the *concept* of industrial safety didn't – we didn't charge for that. Nevertheless, we all focused attention on this aspect because it generated the need as well as being good for society.

I also undertook a course as an Industrial Safety Officer at the technical college in Ultimo. There was a regulatory component to the course, which was especially informative because I had to read the regulations and understand their precise requirements.

By this time I was also reading many books and magazines on the subject, which I would source from all over the place – mostly from America – or from wherever I could. And, of course, I subscribed to the New South Wales Chamber of Manufactures newsletter, which proved to be so crucial when the time came for us to buy the Driza-Bone company.

Taking Gloves Seriously

Later, I designed the classic Driza-Bone coat, but my earliest experiences in designing protective clothing was re-jigging gloves in 'Frank's Workshop'. I was always picking up competitors' gloves to see how they were made.

I found that the best way to design a glove is to listen to customers when they talk about their gloves. They might say, "The fingers are too long", or "They're too thick", or "These gloves don't last long", or "They always bust around the joins around the thumbs", or "They're too uncomfortable..." and so on.

I would ask questions like, "Why are they uncomfortable?" and they might say, "Because there's a big bulk of leather between the first and the second fingers".

Then I would ask myself, "What can we do about that? How can we move that bulk and still make a good looking glove?'

Designing is not too difficult because, generally speaking, we're all a similar size (give or take a few inches) and we're all doing pretty much the same thing when we need industrial protection.

Even though no two hands are exactly the same size – you've got to create one glove to fit most people – and if I design gloves to my size, they will fit most males. My hands are a standard size, which was useful to me for approximate measurements – because if a glove is comfortable on me, it will be comfortable on most other men.

Industrial safety for me was interesting. I enjoyed it and gained great satisfaction out of it. I contributed my fair share to the wellbeing of people in Australia. A lot of people used my equipment designs. Some of our innovations and designs became the standard.

During the 1960s I designed a leather glove that I named the 'R-pattern', and many users adapted our R-pattern as their standard. The governments too used it as their benchmark for all glove manufacturers and suppliers when they sent out tenders. The same procedure applied to many large users such as BHP, and many others.

Keeping Feet Safe

Safety footwear is important, because things are dropped. This can cause injuries to people's feet. It hurts their instep or their toes.

A large number of shoe factories closed down in Australia during the 1960s, leaving only a small number of firms, and it was up to people like us in the safety sector to convince some of these remaining footwear manufacturers to add safety footwear to their lines. The original safety footwear manufacturers used to put a steel cap outside the boot, which was quite strong because the leather underneath gave a bit of added protection even though they were pretty horrible looking things. Eventually, they made better shoes and boots and they put the toecap on the inside.

Nowadays, you can buy safety shoes and you can't see the reinforcements. You can't even detect that they are safety shoes, which denotes the spectacular achievements of recent times.

One of the great things that happened in America after the war was that the insurance firms in America said to companies, "We will reduce your insurance premiums by x per cent per person if you lower your accident rate." Australian insurance firms didn't follow the American example, and I could never understand why.

There's nothing like money to influence people to do things. Apart from the injuries to staff and all the other things that go together in an industrial accident, there's money to be saved by reducing accidents. As I said, it's five times the value of the actual lost time when a person has an accident! If the accident victim loses one day's time, the cost to the company is at least five days. If the victim loses $100 in wages, the company loses $500 in profits.

Stephen

Some years previously, Pat and I had made arrangements for Stephen to attend Newington College in Stanmore. In those days it was difficult to get into any of the private schools and you had to have strong recommendations to get in. Ken France – who I had worked for as a taxi driver – had some influence with the College, and he made it possible for us to enrol Stephen as a full-time boarder. By the time Stephen was old enough to attend Newington, Ken's son Alan was also attending the school, so Stephen had a friend there of the same age when he started.

In 1966, Stephen started at Newington College in fifth class primary. He remained at Newington College until 1972. Stephen enjoyed his early schooling because it gave him an opportunity to mix. He became a full-time boarder except for weekends when he would come home. Stephen is our only child, and we were probably not the best of parents, because Pat and I spent so much time in business-related activities.

Family Life In The 60s

Our working days were eight to 10 hours long, so we didn't have much time to worry about other matters like changes in fashion, popular music or politics, and we were not very politically aware at that stage. We had a business, we had a house, we had a son. We were busy being busy.

Pat and I were squares. We enjoyed movies more than pop music. However, I do remember when the Beatles came on to the scene. To me, their long hair and funny clothes seemed very strange. Before them we had the Beach Boys wearing suits and singing with a smile on their faces, whereas the Beatles jumped around, yahooed and did all sorts of crazy stuff. We didn't *hate* it; we just didn't like the music very much. We preferred to listen to classical music which both Pat and I enjoyed.

Like most Australians in the 60s, we watched TV and listened to the radio, but not too often because our lives were based on what we were able to achieve in the business world more than anything else. We didn't get pleasure from watching *Bewitched* on TV. We got pleasure when we sold 56,000 pairs of gloves to BHP! We got pleasure about doing something good for Hyster Fork Lifts! And we enjoyed many satisfying business experiences and successes. *That's* what gave us pleasure!

The rise of Women's Lib didn't change things much either. As far as meal preparation is concerned, I'm very good at going to the local fish 'n' chips shop, or the local Chinese shop, but as far as other meals are concerned, my cooking never amounted to much. I always did the washing up and Pat has always been the cook.

Driving together to and from work, Pat and I were always talking about our business. Sometimes we would work all day and hardly say a word to each other, except for driving to and from work. We were quite happy to talk business; it wasn't a burden to us. It was something we did because we wanted to.

We would always be the first to arrive and the last to leave, although after the second heart attack in 1967, my doctor told me to stop pushing myself so hard. So Pat and I introduced a personal rule that if it couldn't be done in the ordinary business hours we were doing it wrong. We did our best to apply that principle from then on.

Whatever they say about the 'prosperous' 60s, the economy had its ups and downs. Sure, we made money but we worked hard to do that. We were often the lowest-paid people in the place. We worked for as low as half the wages per hour that we were paying our employees. Many times our combined wages was only equal to one of our staff's wages, so there was no way we could say we were making money in a true sense per hour.

Pat and I both led from the front. We worked *with* the staff. There was nothing the staff could do that either Pat or I couldn't. For example, I could do anything on the sewing machines that any of my machinists could do – and I could do it as well or better. And Pat could do anything an accountant could. For example, she did the profit and loss balance sheet herself; she did not delegate that job. Our auditor came in every three months to check that everything was in order, which it always was.

Messing About In Boats

Pat and I started our boating interests when I was working driving taxis for Ken France who was a champion speed boat driver. On a number of occasions Ken invited me to be a passenger on his boat, which I enjoyed very much. Later, when I worked for John Partridge, he owned a 16-foot half-cabin cruiser and our two families would enjoy going out on this boat.

My main contribution was in keeping it going, doing mechanical and other repairs, so I was able to enjoy both sides of 'messing about in boats'. In the late 50s John sold me his boat. Pat and I then joined the Ku-ring-gai Motor Yacht Club at Cottage Point, the Deep Water Motor Boat Club at East Hills and the Royal Motor Yacht Club at Concord, where I was a flag officer for the club.

I started racing in our first year after buying the boat, and in one year I was successful seven times and won seven trophies. I still give those trophies pride of place at home.

Around about the same time, I went to a boat expo at the Sydney Showground and I was attracted to a stall run by the Volunteer Coast Guard. I was given some information about them, I liked what I read, filled in the application form and joined the branch based at Cottage Point (which was also the headquarters for the Ku-ring-gai Motor Yacht Club).

The Volunteer Coast Guard taught me seamanship in all its facets, including coastal navigation. I'm grateful for the training I received because I was able to enjoy boating much better with that knowledge. After completing a seaman's course, I joined the ranks of the operators who looked after rescue, patrolling and operating the radio service. Every Saturday and Sunday, members were on duty doing those functions.

Over the period of time when I was a member of the Coast Guard, one highlight was the Bridge-to-Bridge Ski Race in which we were the marshals up the river. Other times we manned the security boats and acted as marshals at functions for other organisations.

Mayday, Mayday

One day I had the misfortune of being on duty – working the coast guard radio – when a Mayday call came through. It came from a woman who was sailing a 26-foot Folk sailing boat and had just returned from a trip from England to Australia.

She was sailing into Pittwater, and for some inexplicable reason she finished up on the Barrenjoey Bombora, which was working quite fiercely that day, and she wrecked her boat on the rocks. Her boat was totally destroyed.

A member of our coast guard was on duty and he sent out the relay Mayday. The area of Barrenjoey Pittwater is notorious for having blank spots in radio receiving and dispatch, and no other authority had picked up the Mayday except me and the coast guard at Cottage Point.

I was able to make contact with the Police by landline and the Air Rescue authorities and our own CG18 rescue boat were able to attend to the call and save the lady and her passengers from major injury.

By this time I got to the stage where I had learned enough to be able to run public seamanship courses for the general public at some high schools – St Ives, Turramurra and Asquith. I'd run a three-night lecture and my main function was not to teach, but to organise the publicity and the advertising. At that time I had a rank of Rear Commodore.

In time I resigned as an active member of the Coast Guard but I have continued my interest in boating. Except for a short period during the last few years, I have always owned a boat.

Two-Way Thing

As the decade drew to a close, Armour Safety was expanding. We were gaining in confidence and becoming an important part of the industry, so we decided to open a manufacturing and distribution branch in Newcastle.

By the time 1972 had come around we had two factories – our main one in Sydney where we employed 45 to 60 people, depending on the time of year (our business was a little bit seasonal, so we had highs and lows). And we also had the factory in Newcastle with a maximum of 29 people. We employed around 80 people altogether.

Newcastle is a very parochial city. Take Armour Safety products for example: Novocastrians will buy Newcastle-made gloves ahead of all others because they realise that the machinists are the wives of the people who work for the local firms. It's a double-barrelled shotgun, a two-way thing, and that's understandable.

We opened a factory in Newcastle to capitalise on those market dynamics. We were supplying BHP and the big five.

Marketing

A lot of the items were sold by tender. For example, BHP would put out a tender every two years and everybody chased their business. A lot of other companies were moving into tendering systems, so while we still had to send out reps and sell to industry directly, that was only part of our market. We also supplied the Navy, the Railways, NSW State government stores – who on-supplied them to the

Department of Education and lots of other departments, plus many hundreds of private companies.

We made welding jackets, we made aprons, we made leggings, we made sleeves, we made gloves, we made all sorts of things and there was no foreseeable sign of business slackening. Safety was a fast-growing industry.

And then Gough Whitlam came along and changed the rules.

Part 3

Driza-Bone, The Australian Legend

12. A Raincoat Is Protective Clothing

13. Two Great Men – Emilius Le Roy and Don Pickup

14. Gordon Harman and Driza-Bone

15. Making Driza-Bone Great

16. The Legend

17. Selling The Company

18. Proving A Point

19. The Best Is Yet To Come!

CHAPTER 12

A Raincoat Is Protective Clothing

On 2 December 1972, after 23 years in Opposition, Gough Whitlam swept into office with the famous 'It's Time' campaign. At his first press conference the PM announced the end of conscription and the release of those imprisoned for resisting the draft. This program was a foretaste of big changes that would be forthcoming during the Whitlam years, as his social and business reform program astonished the whole country. This included a recognition of the People's Republic of China, comprehensive universal medical insurance, more spending on education and an increase of social services – so far so good, in a controversial sort of way.

However, the farmers weren't impressed when Whitlam announced that the $12 a tonne superphosphate bounty would cease at the end of the year. And for manufacturers like us, the news was even worse when Whitlam introduced a cut in tariff protection, revalued the Australian dollar and granted women equal pay overnight.

Whitlam did Australians a great service, I suppose, by turning us into a very politically aware people. No other PM before Whitlam had done that. When the Whitlam era ended dramatically in 1975, my final assessment is – I believe he did lots of things that were wrong and also he did lots of things that were right. He certainly shook up the entire nation; that's for sure. We, at Armour Safety didn't have many problems until Mr Whitlam became Prime Minister.

The Whitlam Government had wonderful ideals of what it wanted to achieve but once the Labor team got into power it went crazy. We had changes of regulations every day. We couldn't wait to listen to the evening news to find out what had happened in Parliament that day and what our new obligations were as an employer. Every day was a disaster as far as we were concerned.

Labour-intensive industries were damaged dramatically by Whitlam – and we were labour-intensive. A clicking press operator cuts out six-dozen pairs of gloves per hour – that's 72 dozen per day – and a machinist would make maybe six-dozen pairs of gloves per day – *that's* pretty labour-intensive.

Our first contract with BHP was for 56,000 pairs of gloves. Can you imagine how many people we had to employ to supply that? We had to send the gloves to BHP in wool bales!

Whitlam's policies certainly had a devastating effect on Armour Safety Pty Ltd. We went down from 60 machinists to one-and-a-half machinists and from two factories down to one.

Whitlam Did Us A Service?

Whitlam created a set of circumstances in which we almost lost our business and we would never have taken on Driza-Bone if he hadn't forced us to do so to remain in business.

And if we hadn't taken on Driza-Bone, the company would not exist today.

So in some ways, Whitlam did us a service.

Before Whitlam came on the scene, the previous (McMahon) Liberal Government had already agreed to a policy of equal pay for equal work, which meant men and women would get paid the same wages for doing the same job. The difference between Liberal and Labor policies was that the Liberal Party had decided to phase in the changes over a three-year period. This gave labour-intensive industries a chance to prepare for the change, implement strategies to absorb the extra costs and remain in business.

Whitlam brought it in *overnight!* Not three months, not three years – one day we didn't have it, the next day we did.

Our wage bill went up by 30% overnight, and we had to instantly absorb those costs.

That was pretty hard to take because all our contracts were quoted on the basis of a wage scale that we expected to remain stable for some time. Businesses all over the country had to absorb the same financial shock and many went broke. We, too, would have gone broke if we hadn't made radical changes. The first was to close our Newcastle plant. We sold it and luckily we sold at a profit.

The credit squeeze of 1974 hit manufacturers harder than most. Within two years, half Australia's manufactured goods were produced by only 200 companies.

As if this wasn't bad enough, the manufacturing sector was also hit by Whitlam's tariff policy and non-protectionist stance on Australian-made goods. For example, we had a contract with BHP for half-cotton backed glove at $1.99 a pair – we made very small profits on these in terms of volume. Thanks to Whitlam, BHP could get the same glove from Taiwan or Hong Kong at 98c a pair. Altogether BHP used to buy 300,000 pairs of gloves per year, and if BHP saved $1 per gloves through buying from South-East Asia instead of from us, they saved $300,000. The Big Australian couldn't really afford to ignore that; that's big money.

So overnight BHP cancelled all their domestic contracts. They said, "Your prices are too high compared with overseas." With more than a quarter of a million dollars at stake, you can't blame them for not buying our gloves any more; why should they?

Then, to make all things 'beautiful' again, Whitlam decided to revalue the dollar. So the whole situation became catastrophic for Australian manufacturers. The bankruptcy level was high.

Yes, we sold our factory in Newcastle – we were very lucky we found someone who wanted the premises to make jeans. Our factory was in the inner-industrial suburb of Islington, the building used to be a chemical factory before I bought it. We actually made a profit on the sale of the real estate, which was unusual at that time because Newcastle was feeling the effects of this downturn. So we were very lucky with our real estate, because we really could have been stuck with those premises.

After the decimation of the Australian safety manufacturing sector, just about every item we handled in safety equipment was imported. Most manufacturers had closed down. They just walked away from it; they had nothing left. Safety helmets were imported from Spain, goggles were imported from England, gloves were imported from Taiwan, Hong Kong or China, safety footwear was imported from Argentina, and so it went on.

Armour Safety Pty Ltd almost hit the wall and in so doing nearly joined the 30 other Australian manufacturers who once made industrial safety equipment but have since vanished without a trace.

So one way or another Mr Whitlam forced us to take on Driza-Bone and forced Driza-Bone to become what it is today. I give him credit where credit is not expected. We then decided to try to find some other things to do with Armour Safety Pty Ltd.

Bulletproof

Kevlar is a carbon-fibre material made by Dupont originally for high-speed tyres. It was also useful for reinforcing keel areas in the middle of sailing boats, but it was the fact that Kevlar was good for bulletproof vests, that captured my interest. I wondered whether it would be possible for Armour Safety to move into bulletproof vests, flack jackets and that type of thing.

These ideas appealed to Pat and I because we could still remain in the safety industry, although in a very different way. I had joined the safety industry in 1952 selling goggles and gloves. Twenty years later – still selling safety equipment – I could see myself designing bullet-proof vests and flack jackets.

So I contacted the Australian Kevlar representative and he and I worked out a system where I could buy the Kevlar direct from Dupont and we also figured out how I could get it woven into fabric.

Having sourced the fabric, I went back to my workshop and tried to solve this dilemma: *if a bullet is stopped by a vest, the inertia from the bullet continues through the vest, which can be every bit as dangerous as the bullet.*

At this time a standard vest had a trauma area about the size of a 50c piece. Being such a tiny area, the power wave of the bullet could transfer to the heart, and still cause death. My object was to dissipate the blow by increasing the size of the trauma area, because the *bigger* the effect of the trauma, the less chance it had of travelling through to the heart.

By having the Kevlar material woven into three different grades: one going top to bottom, one going all right, one going all left – and having three layers of Kevlar – we increased the trauma area to the size of a dinner plate. The one that was going both ways was on the outside; its first job was to stop the bullet. The second layer absorbed the pressure and spread it to the right. The third layer absorbed the pressure and spread it to the left. The same logic explains why plywood is so strong.

So after getting a prototype made, I took the vest to the NSW Police Administrative Offices at Redfern. They tested our vests and found out that it stood up well against one of the bigger handguns – the Magnum.

To test it, the police inspector got wet white plaster of Paris and put it into a bag about the same size as the human body. He then sat it on the ground with nothing supporting it, put the bulletproof vest over it and then fired bullets at it from different ranges. Some were fired from a distance of 5-6 feet; some were fired from 20 feet. He used different types of bullets and different sorts of guns, and then measured the effect. The conclusion was that we had indeed increased our trauma area to the size of a dinner plate. The victim would get knocked over and wouldn't be very happy, but would be a lot happier than being dead.

This was the first time this development had occurred in bullet-proof vests anywhere in the world. Nobody told us we could do it, nobody told us we should do it, nobody knew we were doing it – we just did it.

The biggest manufacturer of bulletproof vests was a firm in America. Not even they had worked this one out yet!

Boy, did they soon find out about it though!

Within a month they brought out exactly the same product.

I can only suppose that a member of the Redfern Police passed on the information because I was so proud of myself that I explained everything about the jacket in great detail. Stupid me. I should have just let them test it, then I should have walked out and patented the thing straight away. I was just doing my own thing, learning as I went along and I didn't realise what I had achieved. Unless you're brilliant you don't really know what you are making.

So we lost the race on that one because the American company took the market straight back from us. They already had the name and the reputation and we were too slow to realise the extent of our innovation, so we left bulletproof vests alone and moved on to other possibilities.

Industrial Dry-Cleaning

Pat and I started looking around for other ideas that related to the things we knew – which were factories, leather and safety. There seemed to be an opportunity to do dry-cleaning of industrial leather gloves.

There was a firm in Sutherland that dry-cleaned leather goods, like gloves, aprons and welding jackets. They had it all to themselves. After the reduction of Armour Safety I thought, "Maybe we should try the same."

One of Pat's friends, Molly Smith (a member of the Quota Club of Fairfield which Pat had joined), had a son, David, who operated a dry-cleaning shop in Cabramatta. I asked David, "Could you help us start a plant for the dry-cleaning of gloves?" David knew some people in the trade in Erskineville from whom we were able to buy some secondhand machinery and the necessary things to do the dry-cleaning. We transported the equipment to our Guildford factory where we set up a dry-cleaning establishment with David as a partner.

Our Guildford factory had a chemical sump that had been installed by the previous owner. He used it to wash his fleet of trucks. The sump was designed so the chemicals drained into tanks after which they were carted away by special tankers. The dry cleaning waste needed the same disposal facilities. (The man who bought the factory from us packaged aerosol and no-name supermarket products, and he needed to use this pit too. So this one pit that was build by the trucking business was quite useful for all three businesses.)

We tried dry cleaning for a while but the work was too labour intensive to ever be truly profitable. We weren't making any real money out of it although we had some good contracts – like BHP.

In the end, it was not feasible. It wasn't a real popular job. Staff wouldn't turn up; they didn't like the work. Plus, the time factor was way too high. So Pat and I closed that down and we decided to try other things.

Rainwear Niche

Rainwear had always been a part of our range of safety equipment. I could see that the market for raincoats was not totally filled because there was a niche for over-sized garments for people who were not your 'average' size.

To buy raincoats from overseas an importer would order big runs of 1000 dozen raincoats, of each size. Anyone who wanted extra-extra large couldn't buy them and we could see a significant niche there, so we started making PVC raincoats and we developed quite a steady business.

So we began making extra-large plastic raincoats and trousers, in sizes which the import companies would not handle. Imported ones came in sizes from XOL to SM – a range which covered only five variables – x-small, small, medium, large and extra large, none of which fit big, *big* men. Someone – say, 6 ft 5 in tall, weighing 18-20 stone – would find it impossible to fit into a mass market-sized coat, so we made outsized garments and charged accordingly.

Australian Chamber of Manufactures

As members, once a month we would get an eight-page newsletter from the Australian Chamber of Manufactures. I used to keep them for constant reference because they listed the Labor Government's new regulations. I needed to keep up-to-date with the ever-changing new rules so that I could *dot my i's and cross my t's* correctly and I always read the newsletter closely for that reason. One day I saw an ad on the second last page taking up only an inch of space.

It read: *Wanted: a manufacturer to take over manufacturing rainwear under license or sale.* That's all it said, and then it gave the contact phone number.

Because we weren't doing too badly selling our extra-large sizes, I said to Pat, "Maybe they've got some rainwear manufacturing that we could add to our range." And she agreed that it might be an inexpensive way of increasing our range by doing our own manufacture in other types of rainwear. Who knows where it might lead?

I rang up the number and I found that the company's name was Driza-Bone and the vendor was none other than Gordon Harman.

Coincidentally, I knew Gordon. He was the District Governor of Rotary the year I was invited to join Fairfield Rotary Club.

When we first moved to Smithfield most of the other business folk were small farmers and we didn't get involved with the local people at all until I joined Fairfield Rotary Club in 1967. When you move a factory into a semi-rural area you're very much on your own because the small farmers have lived there for a long time while those bringing industry to the area are new.

I enjoyed our membership and, through Rotary, I became part of the local scene and part of the establishment I was moving in.

Flashback: Rotary Fairfield

There are one-and-a-half-million Rotarians in the world today. It is a growing organisation, twice as big today as when I first joined. We meet weekly, 52 weeks per year, and I still enjoy Rotary. The qualities that make a Rotarian are very ethical. Rotary members are do-gooders trying to do good. I am proud to be called a 'do-gooder' because that's heaps better than being called a 'do-nothing'.

I joined because one day I got a dinner invitation through the mail to attend a meeting at the Fairfield Rotary Club. It was a free night and I thought, "That sounds all right" so I went along and I met all the members who made me feel very welcome. They explained to me a little about what they did as far as what their specific club was about and what Rotary stood for worldwide. And then they invited me to attend an ordinary meeting, which I did. After a couple more meetings they invited me to join and I accepted with pleasure. The club had about 50 members.

By joining the club I automatically made myself available to represent industrial safety equipment in various ways – to the community at large as well as to club members. Although it wasn't planned and it isn't a reason to join the organisation, Rotary contacts became a part of our business circle, which meant that I got to meet a lot of nice people socially. Today, we would call it 'networking'.

Most of the Rotarians I met that night had businesses in the Fairfield area, although (like us) they lived some place else, some from the North Shore, the inner-west and others from the western suburbs. Our club comprised quite a variety of people – taxi drivers, architects, insurance agents, shopkeepers – and now an industrial safety equipment manufacturer!

In those days Rotary was men-only, nowadays it's mixed. I have really enjoyed Rotary. In 1967, Pat had joined Quota, a ladies-only service club, built on the same principles as Rotary. She served as President and was a District Governor Nominee. Stephen joined Rotaract, served as President, and then joined Rotary. Most of the people in these organisations were people just like ourselves who became our friends.

Rotary is made up of districts – zones – and an international head office. In each of the districts there are a number of clubs. I am currently a member of the Ku-ring-gai Rotary Club and our club district has 60 clubs with an average of 40 members per club. There are 2500 members in each district and in Sydney there are three districts.

The first club I joined was the Fairfield Rotary Club in 1967, with an area that starts at Lansdown, and stretches half way to Liverpool, and St John's Park. It is a large area that comprised not only industrialists but also a lot of farmers. I was with them for several years and the most important thing I participated in with them was in 1970 when I organised Rotary-sponsored events to commemorate Captain Cook's coming to Australia in 1770.

My general contribution was my safety expertise, so I put on a Safety Month function in the Fairfield Municipality. In the run up to that month my club members and Pat's Quotarians contacted a total of 80,000 people in a municipality which numbered 120,000 at the time, so we were in direct contact with two-thirds of Fairfield's population.

We had a safety poster competition in every school in the municipality.

We involved lifeguards from surf lifesaving groups from Dee Why to Curl Curl.

We had the local Police checking pushbikes to see if they were okay.

Every kid had a recorded bike number, and they all knew more about their pushbikes afterwards.

We had a lot of publicity on TV and radio.

We got a lot of recognition from the media (which is quite unusual because the media doesn't usually talk about Rotary very much).

We took over local halls and directly influenced a lot of citizens. We had people lecturing to the public about road, home, fire, personal, industrial, swimming, boating and sporting safety.

If you knew of some of the things Rotary does, you'd be very surprised, because as I say they are seldom publicised

Next, Pat's Quota Club started a Miss Safety Queen program – and, ironically, Gough Whitlam was our local member and we had his wife Margaret as one of our judges together with Miss NSW, Suzie Edelman, and the then State Charity Queen. Other celebrities also helped promote this very big project. People donated prizes and a girl was chosen as Miss Safety of the Month. She won a fortnight up at the Gold Coast. We had a lot of cosmetics to give away on that occasion, as well as a lot of other prizes donated by local business houses.

We really got everyone's attention when we organised a big crane that picked up an old Holden, lifted it 80 feet in the air and dropped it nose first. We offered a prize to the person who could most closely estimate the speed it was travelling when it hit the ground. That was quite spectacular.

A District Governor is elected by members of Rotary Clubs to help manage the club and liaise with Rotary International head office in America.

The District Governor the year I joined Rotary was Gordon Harman.

Gordon Harman

I knew about Driza-Bone and I knew about Gordon because of a magazine called *Rotary Down Under* which I received each month. Gordon used to take the back page for an advertisement for Driza-Bone products. I had seen the coats when I was a child growing up on the farms. In those days Driza-Bones were common in the bush, as they are today. I'd even worn Driza-Bones.

Pat and I made arrangements to meet Gordon at his factory in Manly Vale – I believe the building is still standing – and he offered Driza-Bone to us under license or to buy it if that was our preference. Gordon wanted to sell the sewing machines (most of them were pre-war), the proofing tables, the patterns, 5000 pamphlets and his Driza-Bone customer list. That was it.

Pat and I didn't have much money to buy it outright so we decided that we would manufacture it under license for a while instead. This was a long shot because by this time oilskin garments were a no-no in the popularity stakes - the nylon and synthetic raincoats were doing a much better job as far as lightness and smell was concerned. Furthermore, the nylons came in a range of colours whereas in those days Driza-Bones were always khaki brown.

Gordon was in his late-60s. He had married the granddaughter of Driza-Bone's founder, and he owned this family business which was in liquidation, though at the time we had no idea why.

Gordon had lived a full life, especially during the war years when he navigated Lancaster Bombers in England. He had been shot down twice in Europe, after which he came back after the war and took over the Driza-Bone company.

Now it was our turn.

CHAPTER 13

Two Great Men: Emilius Le Roy and Don Pickup

There are eight natural divisions to the Driza-Bone story. Working backwards, they are:

- **The 'new' period – 1989 - to the present.** When Halstead-Belstaff, bought Driza-Bone from us, they sent John Maguire from Britain to run the company in Australia. He now heads a local consortium that bought the company back.

- **Our period – 1974-1989.** We designed the Driza-Bone range of 32 styles, as well as the production techniques and establishing the brand name internationally. We bought the business from Gordon Harman.

- **Gordon Harman – 1945-1972.** Gordon brought back some of the classic design elements and sustained the company for 25 years. He purchased it from his mother-in-law Ivy Pickup and her father T E Pearson. The company was in liquidation from 1972-1974.

- **Don Pickup – 1933-1945.** Don adopted the name, Driza-Bone. Although never a shareholder, as managing director, Don's innovations lay the basis for the 'classic' Driza-Bone coat and injected a sense of enthusiasm for the product. The coat became the focus of Don's attention – not part of a range of canvas goods, as with previous owners.

- **Thomas ('T E') Pearson – 1920-1933.** When T E realised the demand for the coat was constant amongt bush folk, he began to manufacture the oilskin rainwear in Sydney.

- **Edward Le Roy – 1888-1920.** Edward inherited his father's business and continued as a manufacturer of canvas goods. He supplied his 'Roylette' (pronounced Roy-lee-ett) coat to T E's shop, until T E decided to manufacture in Australia.

- **Emilius Le Roy – 1855-1888.** Emilius knew that old sailcloth could be used to make practical rainwear and he was a pioneer because he took that principle *on-shore*. Emilius founded a canvas goods business in Auckland, and started manufacturing treated oilskin coats.

- **Convict days.** The idea of making rainwear out of old sailcloths dates back even beyond the days of Captain Cook. The influences that created the principles behind a Driza-Bone coat are wide and varied. There are probably farmhands on the western plains whose anonymous advice led to various innovations through the years. However, it appears to me that the Australian Driza-Bone tradition begins with Thomas Pickup. Like my great-great-great grandfather, he came out as a convict on one of the early fleets. However, Pickup was a qualified tailer's cutter – thus being one of the founding fathers of the Australian clothing sector today!

Thomas Pickup

When the original white settlers arrived in New South Wales they found no roads, no tracks, no buildings, no structure, nothing. They needed carpenters. They needed quarriers. They needed masons. They needed tree fellers. They needed sawyers. They needed cabinet makers. They needed craftsmen.

And they needed master cutters to make clothing.

The first player in the Driza-Bone story was Thomas Pickup from Blackburn, England, who came to Australia on the *Neptune* in 1818 at the age of 50. His trade was a tailor's cutter or a 'master cutter'. Pickup was found guilty of stealing a yard and a quarter of fabric. In those days everyone in the tailoring trade would take home the small fabric offcuts which they would use to make patchwork quilts, cushions and other household items. All he was doing was taking home his perks!

Pickup was married at the time. When he was shanghaied to Australia, his wife and children had no idea what had happened to him. After a time, they had him certified 'missing presumed dead' and his wife re-married.

Meanwhile, in Old Sydney Town, the Governors were becoming increasingly disenchanted by the harsh treatment given to convicts. Many felt that the better class, or 'skilled' group, of convicts had been hard done by, and Pickup fell among that group. At the end of two years as a 'government servant' assigned to T Clarkson, Thomas Pickup was given a block of land on which he was free to build.

Pickup's talents as a master cutter were used to make clothing for the military and the citizens and who knows perhaps the farmers too.

In a sense that was the beginning of Driza-Bone.

It was certainly the start of the Australian clothing manufacturing sector, and a link between the Pickup family and the origins of the oilskin rainproof coat.

Emilius Le Roy

Emilius Le Roy was born in the Channel Islands on 23 September 1827. He was descended from a French family who emigrated there to escape the French Revolution of 1789.

Le Roy went to London for his professional training under a master sailmaker. (He also found time to play the cello in the London Philharmonic Orchestra.) By the age of 21 Emilius had gained his master's seagoing ticket and spent the next decade at sea.

Emilius travelled on the famous 'windjammer' boats that sailed from England to the southern oceans. They were fast-moving transporters with iron hulls, as many as five masts, and square rigging which required many metres of canvas cloth. The windjammers were constantly awash, even in good weather, as they charged through the seas creating massive waves.

As they have done throughout history, the sailors made their raincoats out of used sailcloth, which gave Emilius a damned good idea! (An early photograph of Emilius Le Roy in his coat shows that the original windjammer rainwear is virtually identical to the current Driza-Bone coat.)

At the age of 25, Le Roy became the master of the *Ilio Mana,* a trading vessel that travelled between Sydney and Auckland. After his ship was damaged in a bad storm, Le Roy settled in Auckland where he established a canvas manufacturing business. Emilius saved enough money to send home to the Channel Islands for his fiancee, Catherine Tabel. The couple married in 1855. Their three sons and five daughters were born in Auckland. In time, their third son, Edward, inherited the business.

Emilius was ceaseless in his expansion plans, expanding into baby needs, flags, tents, tarpaulins, outdoor equipment, marquee hire, camping needs and waterproof clothing. He also developed his oilskin coat with the assistance of his Australian trading contacts.

The original waterproof coat was made out of British standard sailcloth. The warp was a 3-core 60 and the weft was a 2-core 80. It had 180 threads to the square inch – 100 threads on the warp and 80 2-core threads on the weft. At the time it was all made out of Egyptian cotton – grown in Egypt and woven in India under British supervision. Captain Cook probably used the same material for his sails. The use of this particular grade of cotton has remained with the Driza-Bone company ever since. It is the Driza-Bone story's most consistent thread.

Although the first Le Roy oilskin coats are generally thought to have appeared sometime in the late-1890s, Australian visual records pre-date this by at least 20 years. For example, the oilskin coat can be seen in paintings by the Australian Impressionists, as well as in period photographs showing Ned Kelly wearing what appears to be a precursor of the Driza-Bone coat. They were not Driza-Bones; just a water-resistant coat that had taken hold in Australia well before the New Zealand-based Emilius formally added the coat as an adjunct to his canvas goods business.

Apart from running a successful business, Emilius was also gaining a formidable reputation as a violinist. He was a versatile man.

Starting in 1857 as a volunteer, he was also one of the promoters of the City Rifles. He started from the bottom of the ladder and acted as sergeant to his corps in the Waikato war (1863-1864) in which he won a service medal. As a soldier, Emilius fought with the 4th Waikato regiment that also comprised New South Welshmen who were sent to New Zealand to quell the Maori rebellion.

Here, Emilius became friends with the father of T E Pearson from the New South Wales contingency. T E enters our story in the early part of the 20th century. This military friendship grew into a business alliance involving their sons.

As a reward for their contribution to the peace, the Government of New Zealand offered each Australian soldier five acres of land in the town of Hamilton. T E Pearson's father took up the option and stayed.

In 1868 Emilius was promoted to an ensign and later in the same year to lieutenant. He eventually attained the rank of Captain-Commandant, and had the entire command of the six corps forming the Auckland Naval Division.

With military, musical and business interests, Emilius Le Roy spread his talents far and wide. When he died in 1905 he left a legacy which had established his coat in Australia more strongly than in his manufacturing base in Auckland, New Zealand. When his third son, Edward inherited the business, Edward realised that the Australian connection was so strong that he was determined to find a way to capitalise on a market that his father had seeded more than 50 years before.

Edward Le Roy

Until his family friend T E Pearson moved to Australia, Edward Le Roy could not find a way to reclaim the Australian market for the oilskin coat. He spent his energy in developing his canvas goods business while also working on the waterproofing of the canvas coat, which he named the 'Roylette' after 'Le Roy'.

At that time the traditional means of oilproofing cotton was by the application of linseed oil. While very satisfactory for waterproofing, it was found to harden and crack during the long, dry summers. A new oiling process was needed and a unique oiling process was developed which did not crack, did not go hard and stiff, and could allow the garments to be stored for extended periods of time.

Edward hit on the idea of using a mixture of paraffin wax, beeswax and petroleum jelly. Sometimes he even used vaseline in those days.

Australian squatters maintained this coat's reputation from many of the sailors who subsequently left their ships to work on the stations and selections. These garments acquired a reputation for being well made and proved as satisfactory on land as for marine use.

Thomas Edwin Pearson

His wife was actually pregnant with Thomas (known as 'T E') before Pearson went off to fight in New Zealand. When he was rewarded with a land grant, T E's father arranged for his wife to join him and she gave birth within weeks of arriving. Thus, T E became the first white child born in Hamilton, New Zealand.

The Pearson children grew up in Hamilton in a fiercely Protestant family who believed that 'cleanliness is next to godliness'. This catchphrase became the philosophy that encouraged T E's brother to invent, develop and then manufacture the famous Pearson's Sandsoap, which was so successful that he could not keep up with the export demand into Australia. So they decided to cross the Tasman and start manufacturing the soap in Sydney.

T E and his brother found premises in the Sydney suburb of Drummoyne. T E worked his General Store from the front of the building, while his brother made sandsoap out the back. In time, Pearson's Sandsoap started a factory on the waterfront in Rozelle.

On a trip to New Zealand, T E Pearson looked up Edward, the son of his father's friend Emilius from the 4th Waikato Regiment. The upshot of this meeting was that T E decided to stock the 'Roylette' oilskin coat in his Drummoyne General Store. The reputation of the coat was strong and sales were promising, and even though there was a clear market among the farming communities, T E did no more than stock it alongside his range of General Store products. The Pearson brothers appeared to believe that their future success lay in the soap company, which continued to expand until 1911 when, sadly, T E's brother put his foot on a conveyor belt in the factory and lost his leg.

When that happened, T E also became the managing director of Pearson's Sandsoap, which was a much bigger concern. There are many Australian businesspeople who have built empires with far less than what fate had given T E Pearson. In one hand he had the famous Pearson's Sandsoap Company; and in the other, what was to become the Driza-Bone Company. Yet he struggled with both and built on neither.

Nevertheless, no matter how much he neglected it, the 'coat' would simply not go away. Despite years of indifference on his part, during the 1920s when T E was travelling around country Australia he took with him a consignment of waterproof coats and sold the full stock within a week. To his surprise, the coat's reputation was well established among the Australian farmers. All these years T E had remained oblivious to the fact that this had been so since the 1880s!

After consultation with Edward Le Roy, they decided the coats should be manufactured in Sydney as well as in Auckland. After a period of manufacturing the garment at premises in the Rocks, they moved into a small shed in Manly. The Kangaroo Street building still stands where the garment continued to be made until 1933.

Don Pickup

In 1918, Don Pickup came back pretty shaken up from the war. Don had been shot up. His feet had suffered badly and he had shrapnel in his left elbow. His brother had been killed at Gallipoli. He was sent there seven days later without knowing that his brother was already dead.

This returned serviceman met T E's daughter Ivy Pearson at his local Presbyterian Church in Manly. After a brief courtship, they married in 1919.

Don then worked his way up into becoming the head buyer for the prestigious Anthony Horderns store in Sydney. (Anthony Horderns happened to stock the Roylette coat, which his father-in-law imported.) Don remained in Anthony Horderns until early 1933. He left during the Depression because T E stepped aside and offered the managing director's chair to Don. Yet Don never had a personal stake in the family business which he ran.

At that stage there were only two shares in the company – T E Pearson owned one and Ivy Pickup owned the other: father and daughter. But the coat needed an identity of its own; the Roylette name was no longer relevant.

It was Don's brother, the Rev Bob Pickup, who came up with the name at a Pearson/Pickup family get-together. Before the First World War, Rev Bob had been a missionary in New Guinea. After this he became a 'circuit minister' travelling from country town to country town doing all the necessary baptising, marrying and burying before moving on to the next location.

The Rev Bob had just come back from the western part of New South Wales – precisely where Shirley and I were in 1933 – and (the story goes) he told the family that everything was "...dry as a bone, so call it Driza-Bone".

The word 'Driza-Bone' is an Australian abbreviation for the expression 'dry as a bone' which had, for many years, described bones of animals found in the dry, arid centre of Australia. It also means anything that is totally dry. If a raincoat could keep you dry yet not make you perspire, in the sometime torrential rain and cold, you were left 'dry as a bone'.

On 16 August 1933, T E registered the Driza-Bone name under the Trade Mark Act, with the bone as their company logo.

Strangely, the company they formed was named 'E Le Roy Australia Pty Ltd', which remained on the registration papers for another 40 years. Pearson used the Le Roy name in anticipation that Edward would have a continuing financial interest in the business. But Edward never took up his £500 allotment of shares.

Don Pickup was the first person to have a singular interest in the Driza-Bone coat. Don was extremely innovative. In 1933, under his leadership, the first 'Driza-Bone' riding coat was made in Australia and sold here. He wasn't worrying about sandsoap, marquee hire, canvas goods, etc. All he had to focus on was the production of this coat, and he found that he could not keep up with the demand.

During the Second World War, Don Pickup shifted the Driza-Bone factory from Kangaroo Street to Wentworth Street, Manly, where he had five machinists. The Manly factory continued to supply Anthony Horderns and other retailers, and the design shifted from a 'bush coat' to a 'walking' coat.

The Driza-Bone concept had existed for many years and different periods can be identified by slight modifications and changes. For example, the Roylette coat never had the fantail or the cape, although both used the same kind of lining fabric. However, not every Roylette coat was waterproofed. Some early models were made of pure sail-cloth and were waxed on the inside.

During the 1930s the Driza-Bone coats had a flap, but no cape. It was basically a street coat. Another difference is that the classic Driza-Bone has a fantail at the back, whereas the 1930s coat had a diamond cut.

So from 1933 up until the end of the war, they were really making a walking coat, through it had a high cut in the back and straps around the legs for horse riders.

At this time, Driza-Bone was not the only waterproof coat in the marketplace. Eidex was making coats and canvas goods in New Zealand and in Australia there was the 'Stormers' coat. Another coat was made by Evans Pty Ltd in Victoria, who Bradmill took over in time.

As time went by the farmers asked for some changes in the design – first, a long coat for horse riding. The next additions were: the fantail in the back to put it over the saddle, wrist straps to stop the arms getting cold and leg straps to stop the coat from flapping around while on the horse or walking.

Over the years, those who wore the garments refined them to beat the harsh Australian climatic conditions. This culminated in a range of garments with a unique Australian character.

CHAPTER 14

Gordon Harman And Driza-Bone

I n 1944, Gordon Harman returned from the Second World War as a fully qualified instrument maker, and he was full of ideas for making improvements to gyroscopes, pressure gauges, alternators and that type of specialised implements.

Gordon was part of the design team for the gyro-magnetic compass. He also worked with Barnes-Wallace on the Mk 4 automatic pilot that superceded the one made by Sperry. In other words, he had aspirations, and the thought of putting them aside to manufacture oilskin rainwear was the furthest thing from his mind. And then he met Hazel.

While attending the Manly Presbyterian Church, Gordon started smiling nicely at Hazel Pickup, daughter of Don Pickup and Ivy Pearson. Hazel and Gordon's mother were church friends; both also involved in community activities for returned servicemen. And Don, of course, was heading the Driza-Bone family business, having married T E Pearson's daughter, Ivy.

Ivy's daughter Hazel was a flight sergeant in the Decipher Office. She was the section officer and her courtship with Gordon led to marriage – while Gordon pursued avenues for his civilian career as an instrument maker.

Gordon's first involvement with the Driza-Bone company began when his father-in-law was having trouble keeping the sewing machines repaired.

Gordon visited the Manly Vale premises and got 12 of the 15 machines going, so that Don could resume production, which was tough going during and after the war.

"After the war, technicians were as scarce as hen's teeth. This was the trouble with the company – they had the equipment but no one knew what to do with it because there was not a technician in there. So that's when I decided to get interested in Driza-Bone," Gordon says.

Ever the instrument maker, Gordon describes his ultimate contribution to Driza-Bone as "more mechanical than textile".

Major Shareholder

Gordon bought into the company as the major shareholder, though he did not have a controlling interest. Technically, he was buying two shares – that of T E Pearson and his mother-in-law, Ivy. He then created shares for Hazel as well as for her brothers Ray and John Pickup.

Gordon was now the major shareholder in E Le Roy Australia Pty Ltd, the manufacturers of Driza-Bone. The Le Roy name had never been rescinded since the 1930s when T E Pearson had named the company with a view to his friend, Edward Le Roy, coming in as a

50% shareholder – an option that Le Roy never followed through, though his name remained.

The company name remained such until Pat and I bought it from Gordon.

At this stage, all the company shares were held by members of the Pickup family. A meeting was held and 20,000 shares were issued to each of T E's grandchildren. A couple of years later, Gordon formed the Driza-Bone Holding Company Ltd which took over the whole of the shareholding from E Le Roy Australia Pty Ltd. E Le Roy was then declared a non-profit organisation and the whole of the profits went to the staff. This profit-sharing scheme explains why Driza-Bone's price remained stable when everyone else's prices were going up.

While travelling in the southern highlands of New South Wales, Gordon called on Anstiss & Mackay, who had a big store in Wagga as well as a Moss Vale store, both of which stocked his coat. He spoke to Westy Anstiss who alerted Gordon to the fact that the Driza-Bone coat had gone through so many changes that it had ceased to be a proper riding coat and had gradually (an unnecessarily) moved away from his primary market. When Gordon travelled to the northern highlands of New South Wales, he got the same answers.

So Gordon started designing the coat. He actively sought feedback on what a horse rider would need and he generated a lot of interest and recommendations. In the end, Gordon re-drafted the whole coat in a genuine attempt to bring the Driza-Bone coat back to its original outback roots. (For example, before Gordon took on the company, the 1940s coat had a straight seam down the back.)

Manly Vale Factory

Driza-Bone only had five machinists when Gordon took on the Manly Vale factory on the corner of Condamine Street and Kenneth Road. They stayed like that for a couple of years after which the business grew and the company added a second-storey extension that housed another 12 machinists.

In the 1950s they built another factory in Roseberry Street a couple of hundred yards from the first one, where the Driza-Bone company remained for the 20 years or so, until Pat and I bought it.

Under Gordon, Driza-Bone went from strength to strength, until events beyond his control made it impossible for Gordon to make the coat and fulfil his orders.

The Six-Day War

It was the Six-Day War between Israel and Egypt in 1967 that really dislodged Gordon's Driza-Bone enterprise.

The most consistent feature of the development of the Driza-Bone garment over the years has always been its cotton base, which is 'long staple Egyptian cotton'. It is the finest, longest staple cotton that's available anywhere in the world.

It is now grown in other countries but in those days it was unique to Egypt. And like the sailcloth of the previous century, Driza-Bone garments were invariably made of that fabric.

A Driza-Bone coat won't pull out of shape. The fabric is so fine that the threads won't stretch and pull. It has 100 threads per inch running horizontally and 80 threads per inch running vertically. Can you imagine how tight a weave that is? That's why it works! The tighter the weave, the more effective the oil is on the fabric. The less movement in the fabric, the less chance there was of letting water through. That's why ordinary fabrics cannot be used to make a Driza-Bone coat.

When the Six-Day War was over and Egypt lost, they decided to purchase more armaments, which England and America refused to sell them. However, Russia agreed to supply guns, tanks, trucks and munitions, and in return, Egypt agreed to sell *all* its long-stapled cotton to, what was then, the USSR. In this way, Russia cornered the long-stapled cotton market and there was none available anywhere else in the world.

It came on to the market in dribs and drabs, and when it did, it was at very high prices. Poor Gordon couldn't get enough! So his market was gradually dying.

"There was no fine cotton," Gordon says. "We couldn't get any anywhere. We were done. Done like a dinner. The only coats we could manufacture were the heavyweight ones using an ordinary nine-ounce canvas. However, the big market was the riding coat made to Royal Navy specifications out of lightweight sailcloth.

"Health-wise too, I was going downhill. I'd had a heart by-pass operation – and all the usual things – and I thought, "If I continue, I'm only going to be bashing my head against this brick wall here for goodness knows how long."

"So I paid all the staff. Most of them had been with us for donkey's years. Every one of the girls was entitled to their 15 years' long service leave, and coming up around the next corner was equal pay for women. Well, I could see the whole thing was doomed for destruction so we just closed it down. I thought it was time I retired while I still had a few bob. So that was it.

"I couldn't sell the sewing machines for love nor money, no one was interested. But the real estate was worth a reasonable amount, so after two years trying to sell the company, I let go and gracefully withdrew from the textile industry. Driza-Bone had a good name and it was a pity just to let it disappear."

And that's how we came to purchase the Driza-Bone company from Gordon Harman.

Le Roy

As for the Le Roys? They continued to sell camping gear and canvas goods until 1994 from a triple-fronted, two-storey shop in Auckland.

The Le Roy family business also owned a factory from which they made Roylette coats alongside their other lines of canvas goods. In time they let the clothing side run down and they continued to sell camping equipment.

When Gordon went over to see if he could assist Le Roys in manufacturing their coats he found that the Roylette did not even account for 5% of their turnover. The marketplace had been taken over by other players.

Among the market leaders were Driza-Bone in Australia and Eidex in New Zealand.

CHAPTER 15

Making Driza-Bone Great

By 1972 Driza-Bone had been closed down for two years, during which time it was put into voluntary liquidation, although the liquidation hadn't been finalised. That's why Gordon Harman was advertising for someone to continue producing the coat under a license or sale agreement.

So in 1974 Pat and I signed an agreement with Gordon that we would produce Driza-Bone rainwear under license with a view to purchasing the whole enterprise at a later date, though not his factory as we had our own premises. Gordon agreed to come over to the factory as often as necessary, to show us how to make the garments. He also agreed to help organise the import of fabrics from overseas and to assist with introductions to the various shops and re-sellers. In return, we agreed to pay him two and a half per cent license fee on the wholesale price of each garment that we sold.

As far as we were concerned Driza-Bone was a good adjunct to the rest of our product range. It didn't stand out as something that was going to grow. We were still manufacturing other things. At that stage we regarded Driza-Bone as a small but significant part of Armour Safety Pty Ltd.

As soon as we started to work according to the patterns, what we turned out was absolutely ridiculous. Driza-Bones were badly cut, badly made and they were a bad garment as far as comfort was concerned. The oil proofing was fine, but that was the only part that was any good.

Gordon couldn't understand it, so he organised for one of his ex-machinists to come over and show us how to sew them together. Well, she inspected the patterns and was just as puzzled as we were. The sad truth was, during the two-year period that the Pickup family lost control of the manufacture of the Driza-Bone coat, the product had been degraded from being 'the Legend' into the worst coat in town.

I had to start everything afresh. Had the patterns been right, I would not have designed the garment as it is today. I would have primed the sewing machines for maximum performance, I would have improved the efficiency of the company and I'm sure I would have made some modifications to the coats, but the need would not have arisen for me to start from scratch.

We all had to cast our minds into the meaning of this authentic Australian product, and what it evoked in us.

My mind went back to Big Auntie's farm and the old Driza-Bone on the peg. Back to Banjo Paterson's *The Man From Snowy River* poem. Back to Slim Dusty songs. Back to riding horses. Back to the Frank Fisher I used to be when I lived most of my life outdoors. This gave me ideas that helped me design the coat.

Our Son Stephen

Stephen started at Newington College in 6th class in 1967 and was a full boarder. He came home each weekend and brought some other friends with him. Pat and I joined the Parents and Friends Association (P&F) and the Rowing Association. Stephen was very involved with rowing and took on the management of the trainers and their boats. In his last year he was awarded the ONU Service For Rowing Cup, for his contribution to rowing.

Stephen wasn't enjoying 5th form in 1972 and he asked to leave, and we allowed him to. He eventually joined our company, Armour Safety Pty Ltd. Before joining us we insisted he first find employment with some other companies to get some ideas on what being an employee was like. He worked for Kreglingers, a wool selling company, then Olympic Cables Pty Ltd for 12 months, and eventually he asked us for a job. He joined us as an ordinary employee and worked his way up just like anybody else.

Stephen started at the very bottom. He swept the floors, he worked on the clicking presses, he even worked on the sewing machines, and although he was not mechanically minded, he could do small repairs. He never was overpaid; in fact, he was probably underpaid for what he did. It's hard for a boy to have his father as his boss.

But Stephen did well. I was very proud of Stephen and I was astonished that the company who purchased Driza-Bone years later didn't keep him, because by that time he knew everything about the business.

Stephen's real talents were in administration jobs, and he stayed with us right through the heady days of Driza-Bone and worked his way up to the position of Marketing Director and the chief salesman for Australia.

Armour Driza-Bone

When we made the purchase from Gordon, the company name was 'E Le Roy Pty Ltd' which we changed straight away. We decided to include Driza-Bone in our company name so we called it Armour Driza-Bone Pty Ltd. (We took the word 'safety' out of 'Armour Safety'.)

Our receptionists used to answer the phone with the words, "Good morning, Armour Driza-Bone…" and then wait for the person to talk.

One day someone from America rang our office. Our receptionist said, "Good morning, Armour Driza-Bone…?"

And he replied, "Goddam it, ah'm as driza-bone too!"

Eventually the name Armour was dropped out of the name and we were simply known as Driza-Bone.

Sales Investigation

We launched the new business by making a few garments that we showed to the various retail outlets around Sydney, and then Pat and I decided we should visit some of the customers on the list that Gordon had provided. The first town we visited was Orange.

We called on the local AMLF Store, one of the big stock and station agents. I walked in and announced, "Pat and I now make Driza-Bone" and we were welcomed with open arms!

We realised that there were a lot of people on the farms who wanted Driza-Bone coats, but hadn't been able to buy them for more than two years.

(The best alternative to Driza-Bone-style garments was the *Clayborn* coat from Brisbane. Clayborn had learned his trade from Driza-Bone many, many years previously – maybe during the war years or just after the war. He had a little house on stilts in Brisbane where he did all the cutting. The garments were then sewn up by local girls in their own homes. Clayborn only made about 100 coats per month, which is a tiny number considering the marketplace requirements. There were also other small companies, like Melbourne Waterproofs, as well as a couple of others brands that had closed down for the same reason as Gordon.)

From Orange, Pat and I continued driving west to Dubbo and Parkes to call on some of the other country retailers. We were again welcomed with open arms because they could now get their Driza-Bones. The retailers gave us a wonderful reception, and we got a similar sort of welcoming when we went to the other parts of western New South Wales. In fact, our greatest difficulty on our first business trip was getting away from our customers!

They'd put out cups of tea for us, they'd sit around and tell us their Driza-Bone stories, how good the coat was and so on and it became difficult getting away from them. Later on, Stephen experienced the same enthusiastic reception when he, too, called on the same types of customers.

Pat and I weren't quite prepared for this terrific welcome everywhere we went up-country. We had been in this business for some years; in fact, we had tried selling Armour Safety helmets, gloves and other product lines to these farm outlet stores and we had usually received a fairly modest reception. They had never put out the red carpet when we offered the farming community items such as leather harvesters' aprons or harvesters' leggings, but now that we were selling Driza-Bones, there was a sudden change – we became *important* suppliers. So we came back to Sydney with quite a stack of orders.

There was only one problem: we didn't have any material from which to make them. So when we got home, we contacted Gordon and begged, "*Please* get some material from somewhere!"

He chased the world over, but he still couldn't source any long stapled cotton because Russia still had it all locked away and embargoed (and other parts of the world hadn't yet started producing fabrics suitable for oilskin coats). However, Gordon did manage to come to the rescue by using his E Le Roy New Zealand contacts, who supplied us with enough long-stapled cotton to keep our business alive until the worldwide cotton market was restored.

Impact of Driza-Bone

Although Pat and I had added various other lines over the years, the impact of Driza-Bone was much more substantial than anything else we had ever attempted because it affected every aspect of our business. For example, our method of distribution with safety equipment had been to sell direct to end-users, but with Driza-Bone we started dealing with retailers, which was quite different from selling directly into factories.

Furthermore, now that we were 'important' suppliers, by association we had generated much greater selling opportunities for our other Armour Safety lines, which we brought back into the market. Retailers were now buying from me respirators, aprons, gloves, hats and other things now that I was saying that magic word: 'Driza-Bone'. We started making welding jackets and welders' aprons again and we were reaching a much wider market. Furthermore, farmers' methods of production had changed and the farmers were becoming more and more industrialised, now they too had a need for protective safety equipment.

I had always tried to sell as close to the end market as possible. Our sales object always was to cut down the line of 'middle men' between the manufacturer, retailer or end-user. Our reasoning was quite simple: it is because each time you put a product in the hands of a distributor, retailer or re-seller, each of these people gets a margin of profit. We didn't have agents or anything of that nature, we simply sold to the shops ourselves: it was direct selling across the counter. (Driza-Bone today is sold by agents.)

We took no control of the retail price, nor did we care what the mark-ups were. Someone might ask, "Why is Joe's Bargain Bazaar selling his garments $10 cheaper than ours?"

I would reply, "Because that's up to him. Everybody buys from me at the same price. So if he chooses to lower his margin, that's his decision."

Even before our Driza-Bone days, Pat and I were very strict about pricing. We didn't do 'specials' or cheap deals. We never ever fiddled with the price of Armour Driza-Bone products. If a pair of gloves was worth $1, we sold it for $1 – and that was it, no matter who wanted to buy it. We figured out our costs and our mark-up, and then we set our price, love it or hate it. It was fair to me and fair to them. I was not interested in working for nothing.

A Firm Going Mad

I learned to stick to my price from watching others make big mistakes by fiddling with the amount in the hope of gaining a market advantage. This strategy mostly backfired, causing resentment among buyers – though, in one instance at least, the buyer simply played off two price cuts against each other and hurt both.

The episode went like this: in the days when I was with Nicholson Brothers & Lucas, they competed with Protector Safety to sell the same respirator to BHP. A few years later, James North (Nicholson Brothers & Lucas' new name) wanted the BHP respirators business badly so, as an incentive, they offered their cartridges at half price and their respirators at full price. In a competitive move, Protector unknowingly decided to offer their respirators at half price and their cartridges at full price. So BHP bought the half-price cartridges from Nicholson Brothers & Lucas and the half-price respirators from Protector Safety, and neither of them made any money out of it, except of course BHP.

BHP always had super professional buyers and it was a clever buyer who trapped them into this.

So we never believed in selling anything cheaply to gain some other advantage down the track. If a product couldn't stand on its own legs, we'd cut it out, and over the years we did cut out a lot of products.

As a result of our policy, we have never gone through a business quarter, half year or full year, without making a profit. Thanks Pat.

Furthermore, we only ever made one bad investment with somebody who didn't pay us. I never made that mistake again, and I don't want to talk about it.

It was only $4000, but to us it was a significant amount of money at the time.

An Awful Garment

So we started making garments based on the patterns left by the previous makers. For the life of me I couldn't understand why Driza-Bone coats were so hard to make and why they were so bad when they were finished.

It was embarrassing too. We received an order for 20 garments from a Los Angeles retailer (an Australian who used to buy Driza-Bones when he was living here). We were quite thrilled about this 'entry' into the US market.

So we made the 20 garments from the material that Gordon had sourced from E Le Roy New Zealand Pty Ltd, we packed them up, sent them over and then we got a hostile phone call from him, "Call yourselves Driza-Bone? What the **** is this? Not a Driza-Bone that's for sure! It's a heap of ****!"

We said, "Er, excuse me sir, we have followed the patterns with great accuracy."

"I don't want this ****** rubbish!" he yelled, and he sent them all back.

That was very hard to take because we realised that while the world was demanding Driza-Bones, we couldn't meet the market demand because we were stuck with a design that didn't work. And that wasn't the end of it; we had to endure more complaints from various businesses.

Because R M Williams had always been the largest distributor of Driza-Bones, we thought we should call on the company in Adelaide. We arranged a meeting with Dean Williams, the son of R M Williams, who did all the purchasing.

Dean looked over our garments and by the way he shook his head we knew that he was going to rubbish them. In fact, he was quite blunt. Pat and I were very offended. We went away from that meeting feeling quite hurt that he was so disgusted with our garments.

In retrospect, he was entitled to say everything he said. It was an awful garment.

That really forced us into taking some action, so we threw out all the patterns and started to make our own. We could not afford to ignore his advice because when your major customer says, "Go away and only contact us when you've got it right" you've got to come back with the right thing or not come back at all – and we were not going to let R M Williams go.

However, the story has a happy ending. Within a few years, Dean's youngest brother Michael Williams, manager of R M William's Castlereagh Street store, was proudly featured in the newspapers telling everyone about the urban cowboys (or 'milk bar cowboys' as he called them) who purchased Driza-Bones from his store. In fact, he was featured in the newspapers wearing one himself, along with an Akubra hat and – of course – R M Williams boots, a trio which has worked well together as standard Aussie wear throughout the years.

Designing The Coat

When I got back into Frank's Workshop, the first thing I did was to create a stopgap solution to get us out of trouble for the short term. As we were already making plastic raincoats, I transferred the style, sizes and patterns to fulfil existing orders for our first saleable Driza-Bones.

We threw away all the old patterns and we used our own, based on our PVC coats. Straight away we made a better garment. This bought me a little time, which I used to work on the design. Pat wasn't involved in the designing; she has no interest in sewing. She was involved in all aspects of financial management and costs.

We knew what we were trying to do. We wanted to standardise the coat and one problem I perceived was that the Driza-Bone Riding coat was designed for a man riding a horse, but women ride horses too and they also need raincoats. We wanted to make a unisex garment worn equally by males and females.

There were many problems with the existing Driza-Bone pattern, among them:

- **Space.** There was too much space at the back and way too little at the front. It wasn't unisex, which every Driza-Bone has to be. Well-built ladies just couldn't get in there; there wasn't enough space. Anybody with a fair-sized chest would feel constrained at the front.
- **Collar.** It had a very small collar flap on it. A Driza-Bone should have a large flap, to keep the rain out when riding a horse. An oilskin garment does a different job when you're riding a horse than when worn on the street. You need a good waterproof collar if you're galloping into the driving rain.
- **Length.** It was the wrong length and needed lengthening.

In fact, there were so many things wrong that we just couldn't believe it! So I stared hard and long at that coat thinking, "What elements of the pattern should we change to make a better garment?"

Horses and Sleeves

During my childhood, I was brought up on a farm and we rode horses as a matter of course, so I understood the problems a rider might have. From the same period of my life, I was also familiar with Driza-Bone garments. I had worn all sorts of garments and I knew what the requirements were, and I could see straight away that the sleeves were the wrong length, so we lengthened one part of the sleeve and shortened the other.

And that was another fault. Driza-Bone wearers found that when they stretched their arms forwards, the underarm section of the sleeve would tighten up and become uncomfortable because it was pulling against the rest of the side seams, so we made a change to the pattern to overcome that problem. Driza-Bones are not like any other garment, so we could not perceive these problems except by trial and error.

I tried raglan sleeves, I tried let-in sleeves, and I tried sleeves with the seams on the outside. A seam on the outside lets water in and finally I tried seams underneath the arms. It makes sense to put the seam underneath the arm where the water doesn't reach. However, there is a problem with doing that, because a meeting point underneath the arm can be uncomfortable if you put too much bulk there. Nevertheless, our first decision was to use a 'let-in sleeve' which is an inside seam, as opposed to a raglan sleeve, which is on the outside. We found that raglan sleeves used too much material and they were bulky.

Another problem was that the sleeves were too short. Horse riders need slightly longer sleeves because when holding the reins, the sleeves of any garment slide up the arm leaving the hands entirely exposed to the elements – and they can get damn cold.

The sleeve should be long enough to keep the riders' hands warm and short enough not to need constant pulling back because the sleeves are in the way. Therefore, the sleeves of a correctly designed riding coat should reach to the first joint of the third finger when the rider hangs his or her arms straight down by their sides. When the rider puts his or her hands forward to hold the bridle, the sleeves should come back to the middle of the hand, which gives some protection from the elements.

20 Pieces Of Material

There were 32 separate pieces of material in a Driza-Bone raincoat when we purchased the company, and I brought that figure down.

Cutters do not cut each piece out one at a time, they cut up a layer of them. It could be a six-inch or even a 12-inch layer, and they cut it with a vertical knife. (Nowadays they probably use laser cutters.)

I had to keep all these details in my mind, because I had to design something that our machinists could make practically and simply, and that they didn't have to think about too much. You mustn't take away initiative by turning them into robots but you should skill the job down as low as possible to suit the people you've got. Once they've been trained and are comfortable doing the job, their work automatically comes out consistently perfect.

And you have to judge the staff according to their skills. There might be 150 people on the shop floor and I had to spend time with each one to put them in the right job and then they became happy, and happy people produce more! We couldn't arbitrarily delegate the jobs, "You're a machinist, you're a cutter, you're a fitter..."; each person had to be given the job that they were best at doing.

'Classic' Look

I didn't design the new garment by waking up one day and saying, "I've seen the end from the beginning, I am now going to design a new garment." I started by looking at the things that were wrong with the garment that we were making. The 'classic' Driza-Bone look is my design. The appearance of it existed beforehand, though this actual pattern did not.

In fact, the 'caped coat' has been around for centuries. The reason why the Driza-Bone coat was caped was because in the old days, before they began using very fine cotton weave, the cape would only keep out 20%-50% of the rain, so they had put another cape over that, and yet another cape on top of that. By the time they had added five to six layers of capes, those old windjammer sailors of the 1840s had quite a waterproof coat. Meanwhile, the original Driza-Bone company came up with a wonderful formula of oiling the fabrics which brought down the design from multiple capes to a single cape.

Now, a single cape does a number of things. One is that it looks nice. Second, a single cape allows more space under the arms without adding pressure to the seams, which makes it more flexible. And third, it stops the wearer getting cold shoulders.

Have you ever been out on a rainy day with a thin coat on? You get cold shoulders and it's uncomfortable. You also perspire if you are wearing synthetics. You do not get cold shoulders in a Driza-Bone coat because the water bounces off the cape. It's a brilliant concept: if you want to get comfortable in the rain, you wear a caped coat. That's what the farmers had found out. When it's raining and they've got to go outside and work, they can't stand around under a tree. If it's raining, so what?

No More Curves

Not only was the previous collar tab too small, it was also curved like a banana, which had two disadvantages: (1) it let rain in, and (2) a banana-shaped tab takes a long time to make.

It took two layers of material to make a curved collar tab, it was then sewn up all the way round before being turned inside out and the same thing done again other way round –

that's three operations to make one collar tab. I thought, "There must be a better way of doing things!"

So I threw the collar tab away and made it straight. However, I also made a garment that allowed it to be made straight, and it worked.

When I did that, I reduced the cost of the collar tab from 50c to 5c. It's a simple addition: if you've got to put one in every garment at a saving of 45c per garment, the price of the finished garment is significantly reduced.

We did the same with the wrist tabs, leg and underarm straps – we made them all straight.

The storm flap on the front of the coat used to have an angle on it – from high to low. I never could figure out why this was so, and nobody could give me a reason, so I straightened it.

I looked at all these problems and thought, "Well, if an aspect of the old design doesn't improve anything, let's get rid of it. If it's curved, let's just make it square instead." I believe this was a good general rule on the basis that anything that's round or curved is difficult to sew. It's much easier to sew something straight. You just feed it in and let that 31K 20 Singer go as fast as you like, which you can't do when you're dealing with curves.

A Driza-Bone riding coat has four leg, two wrist, one neck and two underarm cape straps. Once upon a time they used to make the straps out of scrap material, but I designed the cuts so that there was no scrap big enough to re-use. Before we changed procedures each strap cost about 50c. By making them all straight, each strap cost only 5c, which is another great price reduction.

Studs

A press stud is two pieces of metal which clip together and fold over so the stud doesn't fall apart. A Driza-Bone riding coat has 17 press studs. Instead of putting a press stud into a wrist strap after it's been sewn into a point, we just folded it into a point and used the press stud to hold it into shape. In this way, one job disappeared altogether and another job was cut to almost zilch.

Unbelievable as it may sound another thing that was wrong with the garment was that the 17 studs didn't even meet in the right places! The previous design had five press studs down the front of the garment with two of them different widths apart, suggesting in theory that 17 was more than necessary.

Once we started to do this it became natural to review every job we did on the rainwear, focusing on the ways and means of improving production or 'making a better mousetrap' as the saying goes. It doesn't sound much, but we were trying to make a garment that a person can genuinely be dry as a bone in, for heaven's sake!

When we first bought the company, it took 140 minutes to make one Driza-Bone riding coat. When we sold the company we were making it in 32 minutes. Think about that – how much money per minute you're paying staff. Imagine the savings in bringing production of each coat down from 140 to 32 minutes, making garments cheaper than Asian imports and keeping jobs in Australia.

Example:

If employees are paid 10 cents per minute
Original time : 140 minutes per coat = *$14.00*
Improved time: 32 minutes per coat = *$ 3.20*
Saving = *$10.80*

Expected retailer reduction
Per coat = *$10.80*
50% margin = *$ 5.40*
Saving = *$16.20*

Isn't that clever?

No it wasn't, it was simply using your brains and dealing with the obvious.

I was also very lucky; I had some very good staff who helped me with the redesign. I didn't think up everything on my own. It was a cooperative effort and in many cases we all made a contribution towards a more efficient and effective way of making a better garment.

Grassroots Understanding

There was nothing a sewer could do on a sewing machine that I couldn't do. I could sew every bit as well as anyone in the place; in fact, I was better than most. I could even embroider on a standard sewing machine. This meant I had a grassroots understanding of what my employees were talking about when they'd talk about a stud, a flap, a length or a shape. I listened to their conversations and responded to their ideas and suggestions. There really was no other way, because – although we didn't know it at the time – Driza-Bone was going through its most significant period of innovation.

If a machinist said, "Mr Fisher, doing it this way is a bit silly. Why do I have to do it like this?"

I'd respond, "I don't know – why do you do it this way?"

She'd reply, "I do it because that's the way it's always been done."

I would then ask, "Is it uncomfortable or something like that?"

And she would reply, "Oh yes, I've got to bend down and pick it up every time I drop one."

It doesn't sound like much, but we solved that problem this way: instead of cutting off every time she finished, she'd leave it joined on.

And when she had a whole stack of them she'd cut them all off click click click and they'd all fall into a box without having to be picked up. (It is difficult to visualise this level of efficiency unless you see the actual functions.)

Employees with a manufacturing problem would ask me about it. They were all genuinely looking for efficiency savers because we had a bonus system – the more they produced, the more money they made. I was very pleased – in fact, I loved it – if they doubled their wages every week, and there were girls who did. If they could produce that well for me, they deserved a slice of the action. That lesson was hammered home to me from personal experience from the days when Clive Nicholson reduced my bonus and with it he reduced my motivation and commitment to his company.

Instead of worrying about the size of the pay packet that I was putting in my employees' pockets, I looked at it this way instead: the more my employee made, the more I made. Second, the best way to attract top-line staff is to offer a bonus system, so that top performers can earn top dollar.

Each machine costs $100 a day whether it is working or not and is used about 23% of the hour and it's all done in bursts – stop/start, stop/start – the other 77% of the time, it does nothing. Now if I can get the same machine running up to 30% or 40% of the hour, it costs no more, so it makes logical sense to get more use out of it because by increasing usage, the price of the machine to the business will come back 50% – or something like that.

This is the type of thinking we had in our business strategy as a means of keeping prices down. It was very important to keep our price below the imports.

Better Buying

A lot of people don't realise that when you're making something, the time it takes is not just the actual physical act of putting it together, it's a whole range of other things too. Pat and I also looked at our hidden costs – for example, the cost of purchase and storage of raw material from which we made the coats. Someone has to buy the raw material, so that person's time and wage becomes part of your actual cost. And that material also has to be stored, stacked and sourced at the appropriate time, so we had to refine that type of thing too.

How long does it take to buy the material to make a Driza-Bone coat? Not very long if you buy in reasonably big quantities; however, if you buy small quantities it becomes very expensive per garment. It takes approximately half an hour to buy the materials for one or 100 Driza-Bones. Then come the delivery costs, the costs of unpacking and the job of putting those materials on the shelves, on top of which there are storage costs. Next, those raw materials have to be carried from the shelves and placed on the cutting table, where they are laid out. Then someone cuts it out according to the pattern, after which someone takes each piece that has been cut out and puts them into their various places on the shelves so the garment makers can go straight to that spot on the right part of the right shelf. Although the actual sewing time might only take 10 minutes per garment, you might spend another 10 minutes doing these other functions to set up the sewers' work.

We did try the Just In Time (JIT) ways of handling raw material stock. It wasn't called JIT in those days. 'Keeping your stock down' was the expression that we used. The idea behind JIT is to keep stocks low, ordering everything 'just in time' as a saving of space and frequent double handling. This close examination of employee time-and-motion wasn't new to us; we had done the same thing with Armour

Safety. We had already worked out the precise costing of our production of gloves, plastic raincoats or other things that we had made previously and we just transferred the same efficiency principles to Driza-Bone. These principles had never been used by the previous managers of Driza-Bone coats, but time-and-motion studies were being used by many efficiency-conscious businesses all over the world as the spiraling cost of wages, rental space and materials forced increased efficiency standards on what was left of Australia's manufacturing sector.

But JIT didn't work for us because we imported a lot of our materials and we could not maintain regularity of supply – most notably the long-stapled cotton fabric.

We had very little power in the timing of supply; for example, in the early days of Driza-Bone, the wharfies caused us big problems. We couldn't rely on them to get a ship unloaded without them causing some difficulties. We always had to keep at least an extra three months stock on our shelves for that reason.

Keeping an extra three months supply of raw materials all the time was part and parcel of our costs. There were times when we had a truck sitting at the wharf waiting for the wharfies to take our load off the ship and we were very worried because if they didn't turn up 'today' we wouldn't be able to get started tomorrow.

As the years rolled on, the wharfies and truck drivers became less militant so we allowed our stock holdings to come down a fair bit.

Pat and I were innovative because we wanted to stay in business, so we were always on the lookout for ways to improve the production as well as the product itself.

I'm Lazy

Cleo Okon was one of my most productive employees. Cleo was a lovely-looking Italian girl who had been sacked from her previous job as a glove machinist because she appeared to be slow at her work – in reality she wasn't slow but 'steady' – but her previous employers couldn't see that.

When she came to me for a job I said, "Show me how good you are on a machine." She sat down and sewed up a glove, and I said, "Fine, start now". And she started.

Like me, she was always looking for the easiest way to do any job, and we used to joke about us being the 'laziest' people on the premises, because both of us were always trying to 'avoid work' by continually improving the system. We would laugh about being 'lazy', but it's really a compliment because a lazy person will always find the easiest way of doing a job. To be quite frank, a dishwasher wasn't invented by a clever woman, it was invented by a lazy man.

Cleo would show me all the short cuts she used to do on the gloves and I really appreciated her efforts as well as her initiative. We remained friends and work-partners for a long time, and at her best she was doubling her wages per week on bonus. Cleo stayed with me for years and years.

Although Cleo made the same gloves as everybody else, she was always our top performer, week in and week out, rain hail or shine. She never worked the machine in fast bursts, so her machines didn't break down as often as everyone else's did. She worked out a way of not stopping by continually feeding in the material. It's exactly the same principle as in the shearing sheds. The shearer works at a steady pace because he doesn't like his machine to break down, and Cleo simply kept the machine running at an even pace chunka-chunka-chunka around the glove, unlike the other employees who worked their machines in stop/start stop/start bursts. Cleo showed me all these principles. She taught me how to get the job done the easiest way.

Cleo is also the mother of the Captain of the Australian Soccer Team.

I remember 'Pauly' as a child.

Other Garments

As business progressed we developed other lines, other products and a good range of coats. When the expression Driza-Bone is used, most people only think of an oilskin riding coat, but Armour Driza-Bone Pty Ltd made and distributed up to 31 styles of garments, plus many colours and types of fabric.

For example, a person riding a motor bike would be happier with a three-quarter length Driza-Bone coat than with a full length one because the full length would get tangled up in the works. The three-quarter coat gives coverage down to the knees, and by the 1970s motor bikes were quite common on farms. Horses were not used nearly as much as when I was working the farms.

As well as three-quarter length coats, we also made half-length coats, battledress jackets, parkas, trousers, s'ouwesters, cyclone coats, Drizalon nylon rainwear, and Driza-para poly cotton rainwear. We also made Armourlon trousers and motorcycle rainwear that we made under license from Belstaff, the English company who eventually came into our lives in a big way.

So we had a great range to meet all requirements.

But as the reputation of our coat grew, protection from wet weather became a much bigger part of our business than the safety equipment side.

CHAPTER 16

The Legend

In the late-70s and early-80s something new was happening in the Australian consciousness. Our Driza-Bone range of products led the field. People started talking about the land Down-Under with pride and Driza-Bone sales started going through the roof.

The Man From Snowy River movie broke attendance records around the world (we supplied 19 coats for that film). The TV soapie *Neighbours* became huge in Britain. Australia II won the America's Cup yacht race. Every Aussie guitarist wanted to own an Australian-made Maton guitar. Aboriginal Dreamtime stories were being published in books. Didgeridoo sales were up. Vegemite got a mention in an international pop song. Aussie icons – like Ayers Rock (Uluru) – appeared on postage stamps, as did the yellow dingo dog. Sales of stuffed koalas to Japanese tourists started going through the roof. People started wearing Red Back Spider designs on their T-shirts. The Australia-Made and tourist campaigns were in full swing, led by Crocodile Dundee.

As soon as overseas entertainers came to Australia they had to have a Driza-Bone coat and a Akubra hat – it became the fashion of the time. We were on the crest of something big – and no one was more surprised than Pat, Stephen and me!

I have never met Slim Dusty, but I knew that he had bought his Driza-Bone from R M Williams and that he proudly wore it to Country Music festivals along with his 'Slim Dusty' Akubra hat. (That design is not on the open market; it is unique to Slim.)

Akubra Hats is owned by the Keir family and manufactured in Kempsey on the north coast of New South Wales. There has never been a direct relationship between Driza-Bone and Akubra, even though there is a sympathetic relationship because the products complement each other. Many times we would find ourselves selling to the same customers, and the same applied to R M Williams boots, especially the classic Craftsman Boot B543.

Our biggest distributor was R M Williams in Adelaide who was our retailer as well as our wholesaler. We gave the company our best price (ie the retailer's discount as well as the wholesaler's discount) which gave them a bit more to play with in their dual capacity.

By this time Stephen did most of the selling in Australia and he was our top salesman. He would go to the equestrian retailers and stock and station stores and tell the people behind the counter all the good things about Driza-Bone coats.

Don't Call Us, We'll Call You

Harking back to the days when leather gloves were my principle line of business, most leather gloves were sold to big industry by tender so, like all the other 30 manufacturers in Australia, I tried to get the business of the Sydney County Council (SCC). It didn't matter how many times I knocked on their door I'd always get told, "Don't call us, we'll call you." The SCC kept buying from their old suppliers, and probably rightly so.

In those days, the SCC supplied electricity to the entire Sydney metropolitan area. (There are other electricity suppliers now – like Prospect County Council, Macquarie County Council and so on.)

The SCC used to have a special glove made for them every year, because they needed a glove that was very flexible when they were handling electrical equipment. It was a hide-palmed glove made out of very soft leather. All the manufacturers of industrial safety equipment wanted to get this business because it was an expensive glove and the SCC were good payers. However, instead of buying from us, the SCC continually made their purchases through older companies with a long-term reputation for being able to supply the type of glove they wanted. They were not easily convinced by Johnny-come-latelys like Frank Fisher and Armour Safety. Can he supply? Can he make sure we're going to get the quality we need? The answer was yes, I could guarantee that, but they weren't prepared to take my guarantee. So we never could get their business.

A short time after we took on Driza-Bone I had to go out to the SCC head office because they had undertaken a new contract which required a certain type of glove which we sold. And so I called on the young man who was doing the buying and looking after these tenders. I explained who I was and he said, "Thank you very much, fill in this form, send in half-a-dozen samples and we'll let you know when we get around to it." That was the treatment I had been used to over the years of trying to snare SCC's business.

As I was leaving I added, "I don't know if you're interested but we've now taken on the manufacture of Driza-Bone coats."

"You've taken on Driza-Bone?" said this young man, suddenly impressed.

"Yes".

"Are you going to manufacture them the same as before?"

I said, "Yes, probably, we've just recently bought it."

He said, "Excuse me for a minute please" then he went away and came back five minutes later and invited me to come into his office where he introduced me to the head buyer at the SCC who I met for the first time that day – after all those years.

It appeared that the SCC had a desperate need for Driza-Bone coats – yes, a *desperate* need – because in the contracts of employment for 'foul weather men' the SCC was required to supply them with a Driza-Bone three-quarter coat and trousers. Not an 'oilskin' garment, not a Clayborn, not an Eidex from New Zealand, not someone else's garment – but a Driza-Bone specifically. The name 'Driza-Bone' was written into their award, and for two years the Sydney County Council had been going off their brain because nobody made Driza-Bone garments any more.

Nobody else could use the name Driza-Bone on their garments. They could say, "It looks like a Driza-Bone" they could say, "It does all the things a Driza-Bone garment does" but they couldn't use the name, so the SCC couldn't buy them because of their contract. There was no alternative. Furthermore, they couldn't substitute them with plastic or synthetic raincoats or anything like that because they are so uncomfortable. Driza-Bone doesn't get hard and stiff when you're up a pole; the plastic garment does. The plastic also gets cold and sticky and they don't breathe, which causes sweat.

This was the problem being faced at head office when Frank Fisher suddenly walked in and said, "Hey, I own the rights to that magic *Driza-Bone* word", which was a most exciting moment.

The buyer ordered 200 coats on the spot and continued to do so every year that we were in business.

The SCC were terrific. I also got some glove business out of that too – *at last!* And they didn't get one cracker of discount; they paid full retail price. They bought not only some of the hide gloves, but they bought a lot of other ones too. Most of their purchases were our good bread and butter lines.

That interaction also made us quite popular with Dalgetys because when the SCC needed Driza-Bone so desperately, we didn't have any material to make them. So I phoned Dalgetys and asked if we could buy back a quantity that I knew they had held in storage for the past two to three years and we bought them back at the price of purchase and we then sold them to the SCC.

Dalgetys was happy because they got rid of some old stock that they didn't think had any value, and the SCC was happy because I gave them the supplies that they had requested so urgently.

Union Man Goes To Jail

There were blue skies on our business horizon. The only cloud came from doing battle with the unions but, after some stress, we even won that too.

We had joined the Chamber of Manufactures during the 1960s primarily because they gave us summaries of the new regulations, new wage adjustments and other government news, and they explained the relevance to our business as these changes came out. Pat and I felt they offered a very comprehensive service. We wanted someone to act as our adjudicator, someone who would operate on our behalf, and the Chamber went to court for us a number of times.

Our first encounter with unions was in the late-60s when we only had Armour Safety Pty Ltd. When we started manufacturing leather gloves we got our factory workers into the Canvas, Leather & Saddlery Award. We were only a small company of 20-30 people and the unions at that particular time were having some trouble with Canvas, Leather & Saddlery Award awards because there were only a small number of people under that classification. There were also one or two big companies that were being affected by the smaller companies like mine and the union wanted to quieten us down a little bit. I do believe there was an occasional 'sooling' of the unions on to some of the little companies.

One day a union representative called to see us. He asked to speak to our staff.

We provided him with a place to meet them, which was in the lunchroom at 12.00 lunchtime. However, he arrived 15 minutes late, around a quarter past 12, and when half past 12 came and lunchtime was over, I suggested that the staff should go back to work. The union man got upset with me about that because he had about half an hour or so of talking that he felt he had to get through.

So he had to come back, and next time he arrived on time. However, again he tried to go past the lunch break and again I told the staff to go back to work on time. This made the union rep very unhappy with us and he made all sorts of public threats that were quite loud and clear to our employees. In fact, a number of our employees who were union members asked me how they could resign from the union because they didn't want to be involved with someone who was so unpleasant. That turned out quite well for Armour Safety as far as I was concerned. (Eventually, the same representative was jailed for embezzlement of union dues.)

However, it was a different story, when our business comprised both leather gloves and coats. We applied to the unions and the Department of Labour and Industry to recognise the gloves under the Canvas, Leather & Saddlery award, and Driza-Bone and our other garments under the Clothing Trade Award. We kept the two award groups separate by dividing our factory into two parts where leather gloves were made in one section and rainwear was made in another.

Luckily for us we had done everything by the book, we made an application to the right people for this variation upon our company's recognised awards, and we had the correct documentation to show that we had informed all the necessary people and had been granted the change in classification. But somehow it all blew up again and because we were a company that had a reputation for not being easy to get along with as far as the union was concerned, it decided to attack. We were cited with a long list of claims of things we had to provide to our staff.

There were certainly some onerous changes to the awards, but the most important part to us was that they wanted to change the awards from Clothing into Canvas, Leather & Saddlery. (Clothing machinists were earning about $8-$10 per week *less* than what Canvas, Leather & Saddlery workers were getting because canvas, leather and saddlery work is heavier.)

Blackmail Or Black Ban

In their wisdom, the unions thought they could probably force the situation because Canvas, Leather & Saddlery came under the 'Miscellaneous Workers Union' awards and the union wanted to get the workers who were making Driza-Bone garments into being paid the same wage as canvas, leather and saddlery workers. We disputed that was the way it should be and after some considerable discussions they decided to declare us black. I could see that the unions were really planning to cause us all sorts of problems.

Being black banned meant that we couldn't get any supplies of raw material and no one would pick up our goods to deliver them to our customers.

After a time, I decided I'd had enough of it, so I applied to my solicitors and asked them to submit a charge of blackmail against the union. Blackmail is a capital offence and a pretty serious charge and I was quite prepared to go for broke because it looked as if they were going to kill me off at any rate.

Second, if they had beaten me it would have set a precedent for all the other clothing companies in Australia. There were many thousands of people working every day at sewing machines in one way or another and, had the unions beaten us, all of them would have had to join the Miscellaneous Workers Union and receive the extra $10-$12 per week extra pay. So it was fairly important that I didn't lose the case – not only for the future of Driza-Bone, but also for the rest of Australia's clothing manufacturers.

My solicitor rang up the union and stated my instructions. He suggested that they might consider their position very rapidly because I was pressing to submit this charge now – not later. As a result, the union certainly did change its tack very sharply.

The secretary of the union flew up from Melbourne to discuss the matter with me. A limousine pulled up outside my factory, with the request from the union secretary, who was in the car, that he would like to take me into town and have a 'discussion'. I was taken to the Labour Club in town, and a short discussion was held with such people as Gough Whitlam, the secretary of the ACTU and myself. In the end they agreed to lift the black ban against me and I withdrew the charge of blackmail. I walked away from there contented. I didn't want to go through a court case. I much preferred the whole thing to die, and it did die.

The third time we clashed with the union was a very similar situation where the union decided we were to be charged as recipients of a log of claims. The trouble was – this one did go to the courts – and fortunately our case was heard by the Justice of the Courts who had actually set the Canvas, Leather & Saddlery award and had also set the Clothing trade awards. This was the Justice who was on the bench the day the unions' claims against us were to be heard.

Of course – having written the law himself – this Justice knew the awards like the back of his hand, and he understood that the whole thing was unfair to me, my company and my staff, and told the unions in no uncertain terms to withdraw the action and the claims. So we won once again. Third time, even luckier!

No 1 Name Product

Like all good things our riding coats have been copied, but we've had very few challengers because Driza-Bones is the No 1 name product.

As a word, 'Driza-Bone®' has also got a generic usage. People don't say, "I'll go and get my raincoat" they say, "I'll get my Driza-Bone" even though it might be a Clayborn, an Eidex, a Stormers, or something else. In England, they do the same with the Macintosh Raincoat Company. They say, "I'll go and get my mac." The Driza-Bone name is now used the same way, even overseas.

We were also efficient with our manufacturing costs. We improved our production so much that we were eating up our costs of raw materials, labour and everything else. During the boom years of the 80s when we had an inflation rate of more than 25% we went three years without putting our prices up. We couldn't make enough garments, we couldn't get enough staff. We were going off our brains! We had the sewing machines, we had the space, we had the orders but we didn't have the people. So we started looking for home sewers – but the Driza-Bone is only an easy garment to make if you've got the right sewing machines; it's not an easy garment to make at home. So home sewers weren't much help. They could do the linings with their overlockers but the rest of the garment had to be sewn up in our factory.

For a while I looked at getting the coats made overseas, so we travelled to Hong Kong and put the word around the different raincoat manufacturers. I gave them a sample and said, "Make me a garment exactly like this for the best price." They came back with a price that was $8 dearer than the ones we were making. They just couldn't match our price, so we were able to keep the market to ourselves on price alone – quite apart from the fact that we had the best name and the best product. So it was tough on our competitors, because the companies who tried to make lookalikes had to use ours as their benchmark price. Yet if they started to sell them at our price, they went broke, because we had such fantastic economies of scale.

Goodwill

During all the years we owned it, Driza-Bone never ran an advertising campaign. We simply didn't need to. The goodwill that came with this 'Australian-made' product was unbelievable!

The Bulletin published a two-page story about our company in July 1985, written by journalist Daphne Guinness. Her article led to other people wanting to know more about Driza-Bone because exposure in a national business magazine with a circulation of 76,000 will naturally generate interest, and more doors started to open.

The *Woman's Day* published a feature in September 1985, written by Dianne Blackwell. I've got hundreds of clippings including the *Daily Telegraph,* the *Sunday Sun, Harpers Bazaar Australia,* the *Sunday Mail,* and even our Rotary publication, all singing the praises of our garment and featuring everyone from Bob Hawke PM to Stephen Fisher, wearing Driza-Bone coats.

The garment was even featured in *Playboy* magazine. They did some beautiful artwork featuring a two-page spread about Driza-Bone garments. The photograph they used was of such quality that it was taken up by our Italian distributor and printed on his pamphlets. The *Playboy* photograph was also picked up by a German distributor and used in the same way. People started ringing us out of the blue to find out all about our product. And then there was Macca!

Macca!

I was an occasional guest on Macca's *Around Australia* program on ABC Radio on Sunday mornings – I miss that nowadays.

It all started one day when I rang him up for an Aussie type of chat – as his listeners do –

and I told him the Driza-Bone story as it was, and we became 'Frank and Macca'.

I didn't have to ring him up after that. If he wanted to know something about Driza-Bone he'd ring and ask me, "What are you doing about such and such?" or "Did you get the coats to northern Queensland...?" And he'd always say something like, "I'm going to the three-day event at Gawler; what do you want me to tell your customers there?" because he talks in a very Aussie manner and he has a devoted following. He would ask very direct questions, all in good fun. Macca's questions (and my answers) always generated not just one enquiry but many enquiries, just through him ringing me up and asking me live-to-air what the hell was going on at Driza-Bone?

Another side of the celebrity equation was that as the garment became more and more fashionable, we had quite a few entertainers offering to plug Driza-Bone if we gave them – say – 100 free coats.

We told them, "No thanks, we don't deal that way. We have an orderly marketing process and we have to look after our retailers."

We always offered a level playing field with our price and we never gave away coats, not even samples. John Farnham paid for his Driza-Bone like everybody else. I didn't offer him any discount price. The same applied to Michael Hutchence.

Liberace, too, paid the full wholesale price for his 17 coats that I'll tell you about soon.

Prince Charles

In November 1977 Prince Charles visited all States in Australia to commemorate the Jubilee Year. While he was in Sydney he played a game of polo at the Liverpool Racetrack on a rainy day, but he must have forgotten that when it rains in Australia, it can be extremely heavy – and he got saturated.

Barbour is an oilskin manufacturer in England with a royal warrant. They made a 'pony coat' that didn't have a cape and was fairly small, designed for riding ponies, not horses. And Prince Charles only had his Barbour coat with him.

He got soaked wearing his short Barbour coat which was never designed to cope with Australian rain, so someone in his party lent him a Driza-Bone coat before he went out to play the next rubber, and Charles was impressed.

So he got into his car with one his polo buddies and went to the local Driza-Bone distributor and bought two coats – one for himself and one for Princess Anne.

We don't know whether Prince Charles continued wearing his Driza-Bone, but after this event Princess Anne has been photographed many times wearing hers.

Princess Anne Pokes Out Her Tongue

Back in the 70s and 80s Princess Anne was in constant turmoil with the media. She pulled faces at newspaper cameramen, she even swore at them and in so doing the press could usually count on a good story if they hung around long enough to annoy the cranky Princess.

In 1978, she went to a big horsy meet in the English drizzle and she put on her Driza-Bone coat, as well as her Driza-Bone sou'wester. She made no effort to look good for the photographers. The Driza-Bone itself was unusual enough but she also pulled her sou'wester down to keep herself dry – and sou'westers are not a pretty looking garment at their best. Furthermore, the way she wrapped her scarf around her neck didn't add much either. So Princess Anne was quite a sight.

She looked very different and nothing like a princess.

When the paparazzi caught sight of her walking around in these odd-looking garments they took lots of pictures of her. Then the press began to gather and started asking her to justify her appearance. She coolly replied, "The best garment to wear in inclement weather is my Driza-Bone from Australia" then she spun around, took one of the photographers to task and poked her tongue out at him just as he snapped her photograph.

That became the front-page picture in the magazine section of the *London Times* and went on to win the Best Picture of the Year. It was picked up all over the world: Princess Anne in her Driza-Bone coat, poking out her tongue at the media!

Princess Anne honestly believed the garment was the best one for its use, and she used it many, many times. We saw many photographs of her in her Driza-Bone coat at horsy meetings, and she was quite free in telling people what it was.

So that was great publicity, because Princess Anne is probably one of the best-known equestrians in the world and Driza-Bone garments got a wonderful kickalong because of her.

Shortly afterwards, the Driza-Bone coat became the fashion fad of the Sloane Ranger Set.

Most of the people who belonged to the Sloane Ranger Set lived in the Kensington-Chelsea part of London where Sloane Square is located right in the centre.

More Royals

When I was in London on a selling mission in 1981, I had a 10.30 am appointment in Kensington with a potential buyer but as I was running early, I decided to have a cup of tea while I was waiting. I walked into a High Street café where there was only one available booth and I asked the waitress, "Can I sit in that booth?"

She replied, "Sorry sir, it's booked."

"I only want a cup of tea and a piece of cake," I meekly answered.

She said, "Okay – seeing as the other people haven't arrived yet, you won't be long will you?"

"Oh no, only 10-15 minutes."

"Have the seat then."

Three minutes later the people who had booked the table arrived and the waitress said to them in a slightly embarrassed tone, "I'm so sorry, but I let that gentleman use your table to have a cup of tea. I'm sure he'll only be a couple more minutes."

Then one of the girls asked, "Where is he from?"

The waitress replied, "Going by his accent, I think he's from Australia."

They were only a couple of metres from me and I could hear every word they were saying, so I turned around and said, "That's right, I'm Australian."

"Oh!"

One of the girls – who was in her early 20s – came over to me and said, "Can I talk to you?"

"How can I say no, after pinching your table?" I replied. So she sat beside me.

She was accompanied by two older women, maybe a mother and an aunt. And then this girl introduced herself and said, "My name is…" and for the life of me, I can't remember her name!

I said, "Pleased to meet you?"

She said, "What are you doing over here?"

"My business is selling raincoats."

"Raincoats from Australia? Are they Driza-Bones?"

I could have fallen over with shock!

When I replied, "Yes" it was her turn to fall over with shock!

After talking about Driza-Bones for a while she told me that, as Princess Di was destined for marriage to Prince Charles, she was taking on Princess Di's job at the Young England Kindergarten in Kensington. She was extremely interested in Driza-Bones because all of her set wanted to know where they could purchase Driza-Bone coats!

She obviously knew Princess Di and Prince Charles quite well and after that I think she probably sold a lot of Driza-Bone coats to her friends. She certainly did a good selling job on me!

This brief interaction made me really feel that we were in the right place with our garments – right here in Kensington. I talked to her for an hour, and although I was late for my appointment, under the circumstances I thought it was a great investment of time.

Quite a number of members of the Royal family bought Driza-Bone garments. I'd wager the Queen has probably got one too. Harrods sold them for £115.

Range Roverish

The Range Rover Set cast a wider net than I thought. When I was back in Australia, Range Rover released a new range of cars and Princess Anne's husband, Captain Mark Phillips, was employed by Range Rover to promote them. They decided to accompany its release with a range of accessories – Range Rover watches, Range Rover outdoor chairs, Range Rover hats, and significantly for us, they also offered Range Rover coats and decided these should be Driza-Bones.

We made coats that carried the words, "Driza-Bone made exclusively for Range Rover", and we continued this arrangement with Range Rover for some years.

They chose British Racing Green for the colour and we spent time designing the Range Rover garment – not that we changed anything radically but we just made it slightly more Range Roverish.

When Captain Mark Phillips came out to publicise this new range, we were invited as special guests to a place in the backblocks of Sydney where the Range Rover Company took over a few acres of land and test-drove their cars around mud tracks. Mark Phillips and some other name equestrians took part in this rally.

I went for a ride with Mark Phillips in a Range Rover that he was driving until he got stuck in the mud.

He got so bogged that he couldn't move because all four wheels were suspended off the ground. He seemed so distraught.

As I had my video camera with me, I quickly hopped out of the car and captured a very frustrated royal husband on video. In the end a rope was affixed to his Range Rover and it was pulled out, to his relief.

Better, Better, Better

I was always trying to improve the garment to make it better – a better finish, a better hang, new material and other possibilities – so every time I went overseas I always took a suitcase full of garments. I also brought them all back again, because I never gave anything away.

By this time I used to travel overseas three or four times a year. However, whenever I went overseas, the weather always turned fine, so there was no reason to wear the raincoat.

The longest I have ever worn a Driza-Bone coat overseas in the rain was 20 minutes. Pat was with me at the time. It was raining for a change, so we both wore our Driza-Bone coats from Harrods to Green Park in London.

That's the only time I ever wore it in the rain.

Malcolm Fraser's Bodyguards

Virtually every overseas Government member who has visited Australia from 1980 onwards has been presented by Australia with a Driza-Bone coat (and an Akubra hat) as part of their image of Australia – examples are the former Japanese Prime Minister Noboru Takeshita and former British Prime Minister, Margaret Thatcher. And many Australian politicians like to be seen dressed in this way, especially to attract country voters at election time. I've got pictures of Bob Hawke in a Driza-Bone coat. I've got pictures of other parliamentarians too; however, Malcolm Fraser turned out to be a real classic!

Malcolm Fraser, who owned a pretty big station in Hamilton in the western part of Victoria, was the Prime Minister of Australia from 1975-1983, and is often seen in his Driza-Bone three-quarter coat. Fraser is 6 ft 6 in tall – so the biggest garment we made was just big enough for him. And he *loved* it!

Pat and I were invited to a function at Sydney's Randwick Race-course where Malcolm and his wife Tammy were guests of honour. There was whole bunch of people inside and probably 200 people outside, and Malcolm and Tammy were moving through the crowd, shaking hands and saying hello to everybody. We were among the crowd of people waiting for him to come to where we were standing so that we could shake his hand and say the usual platitudes. But somehow or other Tammy reached us before he did and she introduced herself with the words, "Tammy Fraser, and your name is who?"

Pat said, "My name is Pat Fisher."

Tammy Fraser said, "Oh ! Oh hi! How do you do?"

Then Pat added, "By the way Mrs Fraser, do you know what sort of coat Mr Fraser is often seen wearing when he goes fishing or when he's on his motor bike?"

Mrs Fraser said, "Of course, that's his Driza-Bone."

"That's wonderful, because we are Mr and Mrs Driza-Bone. How long has he had it?"

"Twenty three years – when he's at home you can't get him out of it."

"Is the garment still all right?"

"I think so."

"Well, if it starts to leak you know that you can send it back and we will reproof it for you?"

"Oh right-o," she said. "That'll be a good idea. Do you have a card?"

Pat's cards were in her handbag, so it was easier for me to pull one out of my inside pocket.

I reached for my wallet just as Pat reached for her handbag.

Suddenly their minders (who we hadn't noticed before) sprang into life when my hand dived into my pocket and Pat's dived into her handbag. Two super-strong bully boys interceded, thinking we might be getting out our guns. Tammy was also a bit agitated because she wanted a pen or pencil to write down the details, but she dared not asked us to get anything else. Not now that we were surrounded by these hefty bodyguards!

We chatted for a while about how good the security was, then Malcolm came across and she introduced us to him as 'Mr and Mrs Driza-Bone'.

The following day his private secretary rang us at work, "I believe you offered to do up his coat if it was leaking?"

"Yes."

He then said, "Malcolm Fraser was quite pleased with your offer and would like to know if you'd like a photograph of him wearing his Driza-Bone coat?"

"We'd love one!"

"You can't use it for advertising or publicity," we were told. "It's just a keepsake to acknowledge that the Prime Minister of Australia is wearing a Driza-Bone coat."

"That would be lovely," I said. "And would he also sign it?" So he autographed it. Fraser was a significant man in his time and he was happy to wear his Driza-Bone coat and to recognise us as its manufacturers.

Liberace's 17 Driza-Bones

Many internationally famous people who bought Driza-Bones wanted to know the story behind the name. On our Driza-Bone swing ticket we had a thumbnail sketch of the history of the original Driza-Bone Stockman's coat, and we would give a copy of that story to people who would ask about our history.

But the story behind Liberace is worthwhile telling. It started with a phone call from his Australian representative who asked if he could come out to our factory to purchase some Driza-Bone coats for Liberace and his entourage.

Accompanied by some members of his road crew, Liberace's representative (who was also his bodyguard) came to our Guildford factory later that day. He was built like a sumo wrestler and was the biggest man we'd ever been asked to make a Driza-Bone coat for, so we had to make some adjustments for him. Although his crew was happy to have their coats in traditional brown, Liberace wanted something different.

At the time I had been experimenting with oiling the various colours and I had established that the special oils that are used to make Driza-Bone products completely waterproof also makes fabrics go dull. I had, in Frank's Workshop, a red fabric which had turned into maroon after having been oiled, a white which turned into khaki grey, a yellow which had turned into puce, and a light blue which had turned into an oily yellow-blue. I showed these samples to the representatives, who asked for a sample of each to take back to show Liberace who liked the puce and the maroon. His final order was for 17 standard riding coats, two XXX large – for, guess who? – plus one puce and one maroon for himself.

When news of the order got around the factory, everyone wanted to make some part of Liberace's coats. Little notes were sewn into the seams and into the pockets, wishing him a warm welcome to Australia. Liberace was quite touched by this gesture from the staff and he sent his personal thanks to everyone, as well as a pre-released record and an invitation to attend his concert as his guests the next time he came out to Australia.

Unfortunately, he died before he made his next visit, but the staff got a great deal of pleasure from his use of Driza-Bones and we all thought of him with affection.

Over the years 1974-1989 many famous persons bought Driza-Bones and whenever we heard of such purchases our staff was informed and joined in the pleasure that we all had in making Australia better known.

Dun and Bradstreet

Very early in our business life we registered ourselves with Dun and Bradstreet, the business information and auditing company.

We did this because we wanted a marketplace rating. Dun and Bradstreet's ratings are based on turnover and credit.

Being listed meant that our records were available to anyone, which led to a lot of people making contact with us.

Move To Queensland

During 1986 Driza-Bone had clearly outgrown our Guildford premises. We needed to move somewhere where we'd get more space and a better staff situation, even if it meant moving interstate. We didn't have enough staff to meet all our orders and we were knocking back too much business. The lookalikes were moving in and pinching bits of the market that we had generated. We were losing our way and leaving a great gap in the marketplace because we could not keep up with the demand.

After looking around, we finally found a place in Queensland in a little suburb called Eagleby, near Beenleigh. The building had been a hotel-tavern that had been built by Russ Hinze, the famous long-term Queensland Minister for Gaming. He was well known all over

Australia: big fat Hinze. He had persuaded somebody into construct-ing this beautiful building in a place he thought was going to grow, but it just didn't. Nobody came to the tavern, which closed down after a few months.

It was a lovely looking building, modern, with air conditioning, lots of lights, lots of toilet facilities, and because it had been a tavern, there was ample parking, storage facilities and all the things we wanted. If Pat, Stephen and I had set out to build a Driza-Bone factory, we couldn't have built something as ideal for that price.

However, by the time we saw it, it was no longer a tavern. It had no goodwill. It was only worth its real estate value. They couldn't get a buyer for it because it was big and awkward to sell, so from our point of view, it was well priced. In fact, we were able to move to Queensland and buy the Eagleby site for only slightly more than we'd received from the sale of our Sydney premises. It was ideal for what we wanted. Here we had 26,000 sq ft of space on 10 acres of land in the middle of one of the most depressed parts of Queensland. We could offer employment to 150 people. The Driza-Bone Company is located there, to this day.

We employed all this staff and increased our production by 200% within three to four months and our annual output reached a peak of 100,000 riding coats in 1988.

Having decided that we were going to buy the place we made the decision to start manufacturing the day after Australia Day, which was 27 January 1987, and we went straight into 'move mode' after we achieved that goal. We were totally out of Sydney by my birthday, 3 April.

It's amazing how well it worked out, except that it was difficult for Pat and I because we had to live apart in two states during the move, but we were able to exist without each other. Just.

The move from Sydney was a very big deal, we had to move key staff, we had to get replacement staff, we had to take machinery to Queensland and get it operational and we had to take the whole business to Eagleby, with all its paperwork. Moving house is a big job – but multiply that by at least 1000 and that's what we are talking about, Yet it all came together without any major hitches.

I was an optimist. I could see things were happening with the minimum of fuss.

Pat's Illness

Not long after the big move to Queensland Pat had a funny turn. The doctor diagnosed what he thought was a heart attack, which the tests did not confirm. She had three other attacks and then a type of stroke. Pat was hospitalised for nine days and told to take up yoga for relaxation. Nevertheless Pat's condition worsened.

A Rotary friend recommended a Brisbane doctor who put Pat in the Mater Hospital for another 10 days where she underwent various tests, which found that she had a hole in her eardrum, a growth in the middle ear and other problems.

Because the infections had been left unchecked, they set up an angina of the nervous system. Pat had an operation in October 1987 and within six months the growth returned and the entire operation had to be done again. This was eventually done in March 1989 after we finished our working lives.

World Fair Expo

Our move to Brisbane in 1986 was great timing because 1988 was the year when the City of Brisbane held the World Fair Expo and invited every nation to put up an exhibit in the Brisbane Exhibition Grounds. Eighty countries had a presence there, and on the last day every member of each country's delegation as well as their assistant were given a Driza-Bone coat as a memento of their Australian trip. (They were also given an Akubra hat.) Stephen and I had to drive up to Brisbane from the Gold Coast with a great truckload of 200 coats. All the delegates were photographed in their Driza-Bone coats.

The organisers of the Brisbane Expo paid full wholesale price for our garments. We charged the same price as to any retailer.

When I sold leather gloves, if someone said to me, "I'd like to try your gloves out", I would say, "Fine, I will sell you a dozen pair. I promise that if they don't do the job I will let you have them free, but in the meantime here's an invoice for them."

In all the years I did that I never had one pair of gloves ever sent back as not living up to its promises. The same is true of Driza-Bone coats.

Ronald Reagan, I Don't Believe It!

Pat and I went to the office early one morning to get a little peace, to catch up on figures.

We were in my office when the phone rang. The switchboard was switched off except for a direct line into my phone and Pat's. Pat answered in her best switchboard operator voice, and this lady asked to speak to someone who could give her details about the Driza-Bone Riding coat – Pat passed the call over to me.

The caller wanted to know how heavy the garments were, how long would it remain waterproof, if they would they fit both a man and a woman, what colours were available, did we have stocks in America and what price were they?

I explained that the man's coat weighed approximately three pounds, that we had brown and navy blue in stock, and that the price could be ascertained from our American distributor.

The lady then asked me if I could send a pamphlet and details to her. I asked her name and she replied that she was calling on behalf of the President of the United States of America, Ronald Reagan.

As I had some friends who sometimes pretended to be potential customers, I assumed this was a joke call, so I replied, "Oh, and my name's Mickey Mouse!"

"No, no" she said. And after some persuasion she convinced me that it was a genuine call, and Ronald and Nancy Reagan got their coats.

One Price For All

Over the years, Presidents, Prime Ministers, pop stars and all sorts of visitors to Australia have been given a Driza-Bone coat almost as mandatory of part of their Australian experience.

Customers included Elders IXL, R M Williams, Dalgetys and hundreds of other domestic retailers. Overseas, Neiman Marcus, Bloomingdales and many other big stores in America and throughout the world sold out as soon as the coats came into their stores.

Lady Tryon (or 'Kanga' as her friend Prince Charles called her) sold Driza-Bones from her exclusive shop in Beauchamp Place in London.

Singers Olivia Newton-John, John Denver and *Perfect Match* host Greg Evans, and pop star politician Peter Garrett; actors, Kirk Douglas, Jack Thompson and Tom Burlinson, as well as cricketers, Steve Waugh, Geoff Lawson and Mark Taylor, are all Driza-Bone wearers. And the coat was featured in the TV productions *A Town Like Alice, 1915* and *High Country*.

Throughout all this we did not deviate from our one-price-for-all policy.

Giving away product is very expensive. I've never believed that a business can afford to give away its product.

CHAPTER 17

Selling The Company

By the 1980s, it was clear to everyone in the world that Driza-Bone was unique in every way. We had thrown away the rulebook and the traditional ideas about how a garment should be put together. We didn't make garments like other manufacturers. This made our garments different and because of that we were much faster, and therefore more internationally competitive, which led to us granting licenses to certain English manufacturers.

All I set out to do was make coats in the cheapest and best possible way, so I had to introduce the ideas that I outlined in chapter 15 to our licensed manufacturers in England.

Their factories had been making garments for years in the traditional way. The way I had organised things must have seemed quite strange to them.

For example, a normal coat sleeve is made from two pieces of fabric, yet we made our sleeves out of one piece of fabric. By doing this we eliminated one row of stitches. Why did they make them out of two pieces of fabric? I don't know and I don't care.

Another example: in the old days the riding coat was made from 33 pieces of fabric:

	Old days (pieces)	New ways (pieces)
• Body:	4	4
• Sleeves:	8	4
• Collar:	2	2
• Flap:	2	1
• Patches:	2	2
• Straps:	10	7
• Cape:	2	1
• Collar tab:	2	1
• Failtail:	1	1

To save time, I wanted my staff to work with fewer pieces, and by using more folds, I brought the number down from 33 to 23 pieces, which were less to cut as well as sew. This made our manufacturing costs much more economical.

Using these same principles on all our range of garments, our prices remained below rival imported products.

The Driza-Bone range of style and colours meant that we made over 30 variations of the coat. The sizes were: child 0 – 1 – 2 and adult: 2 – 3 – 4 – 5 – 6 – 7 – 8 – 9 – 10 – 11. Each size came in traditional oilskin brown, navy, black, British Green, red and in dry-oiled brown, blue, green, Aussie Gold and yellow. Furthermore, the oilskins came in three weights: X super fine, super fine and heavy weight.

Francis Webster

As Driza-Bone's international reputation grew, we decided to find a way to produce some of our garments in Great Britain, so we took the question to our supplier of raw material, the Francis Webster Company Ltd in Arbroath, Scotland.

Francis Webster was a very old cloth manufacturer. Its history goes back to the 18th century and their initial line of business was making fabrics for the British Admiralty in the days when the sailing ships ruled the oceans. Francis Webster had a reputation for making first-class material. The company had previously supplied the old Driza-Bone company and understood our requirements. We built a relationship with the company, and when we decided we wanted to set up a license to manufacture Driza-Bone in the UK we turned to Francis Webster for advice.

I asked the managing director if he knew of a smallish company that would consider making Driza-Bone garments under license in the UK. Ideally, we wanted to attract a licensee who had some experience with oilskin fabrics – or 'waxed' materials – as they call them in England.

They put me in contact with a fellow called Derek who had a company called Chatsworth in a small town in the midlands of England. I called on him and he said "yes". He and I should spend some time training his small staff to manufacture our garments. He was not only going to be the Driza-Bone manufacturer; he was also going to be supplier and stockist of the UK garments so I introduced him to our distributors. By this time we had a number of distributors. One was Lily Whites of Piccadilly, another was a firm called Casual Riding near Richmond, an outer-London suburb, as well as a couple of other firms.

Trophies

When we first appointed Casual Riding Co as our Driza-Bone distributor, we supplied them with a range of our oilskin garments from Australia which Casual Riding displayed in pride of place at the prestigious British Equestrian Trade Association (BETA) Fair. Our riding coat was judged the best new British product of the year!

BETA presented Driza-Bone (Casual Riding) with a certificate and a trophy featuring a sculpted horse's head. Unfortunately, three years later, our Australian office was broken into and the trophy was stolen, which was a real disappointment.

Many times in later years our Driza-Bone coats and our Driza-Bone business have won international, domestic and regional awards. In fact, there are too many to list.

Each year every Australian State runs a major Small Industry Award and the winners compete for the Federal Small Business Award.

Armour-Driza-Bone Pty Ltd won the 1988 Queensland Award and rumours told us that Armour-Driza-Bone were front runners for the Federal Award.

However, in the Bicentennial Year of 1988, the Federal Small Business Minister, Barry Jones MP, decided not to present an overall national winner for the year ... so we'll never know!

From Chatsworth To Belstaff

After spending some time with Derek and Chatsworth, we reached a point where he could produce the garments without any further supervision. So, after having been in the UK for three weeks, I came back home.

I had only been home two weeks when I received a most unexpected phone call from Derek who said, "We've gone broke, we're out of business and we can no longer manufacture your coats."

I don't know the details, but his money ran out and he was forced to close his doors. So I jumped on a plane and flew straight over to England to ascertain the situation. Regrettably, everything he had said was true, Derek's company was unsustainable and he couldn't do any more business with us. So I had to start from the beginning again in Great Britain.

My first job was to get hold of everything in his company's possession that related to Driza-Bone, because by this time we had significant interest from other outlets. I now had to begin another search for a licensee that, fortunately, was not too difficult. I tracked down a small manufacturer of outdoor goods – a Spanish fellow in Peterborough, also in the Midlands. We came to an arrangement and he made our garments for the next four months.

Belstaff Pty Ltd

For quite some time we had been the licensed Australian manufacturer of Belstaff Motor Cycle Clothing products. Rather than having to pay the 30%-40% duty (or whatever the rate was at the time) on the imported finished product, Belstaff found that it was more cost effective to send the raw materials to Australia and let us make them.

The Government actually made a right decision in favour of Australian manufacturers. This is one of the few times when Australia's customs arrangements worked sensibly for Australian manufacturers. They must have made a mistake!

Belstaff was a well-known English company that manufactured 'Black Wax (oilskin)' motorcycle coats, trousers and motorcycle mitts. They were the best-known brand name for that type of product in England as well as in Australia. They were selling their products to all parts of the world including the United States, but their biggest market was the United Kingdom and Europe. They had distributors and agents who sold their products to the EEC, the EFTA countries and elsewhere.

I eventually got around to talking to the management about manufacturing our garments in the UK. Belstaff acknowledged that our garments were very well received – above all, they were Australian and completely different from anything available anywhere in the United Kingdom, the continent and, indeed, the world.

Yes they said they would consider the possibilities and after some further consultation, they decided to take over all the production and become our sole licensed UK manufacturer.

At this time Belstaff had four factories. Most of their oilskin garments were made at Silverton where Belstaff employed 400 people in a big factory. I was requested by the managing director to visit their plant to explain the manufacturing procedures for Driza-Bone garments to the factory management.

Always in the background was a fellow called John Maguire, the production director of Belstaff International Ltd.

As I was dealing directly with the managing director of the Belstaff company, I hadn't had any need to meet Mr Maguire. However, one day when I walked through the company's administrative offices, he came into the room and we were introduced to each other in a pretty casual way. (I was introduced to people all the time, so I didn't know what his position was, nor did I care very much.)

Someone said, "This is Frank Fisher from Australia and this is John Maguire." We shook hands and then he left.

I had nothing else to do with him, until later.

Two-Way Deal

I returned to the factory premises and spent time with the factory staff instructing them how to make Driza-Bone garments according to our specifications and efficiency standards. I showed them how to make a coat in 32 minutes, using 23 pieces of material. I showed them how to save time by folding the 17 press studs, and how to make sleeves out of one piece of fabric. Boy, oh boy, I got them motivated!

By this time our arrangements with Belstaff had evolved into a two-way deal, I had been making Belstaff garments in our Queensland factory for the Australian market under license and now they made our Driza-Bone coat under license in the UK.

During the early to mid-1980s my company and theirs worked closely together which meant I had to make regular trips, three or four times a year, to the United Kingdom and other places around the world on behalf of the joint-businesses.

And then one day things began to change. They told me that their managing director was being replaced by David Brocklehurst, who was formerly the company accountant. David Brocklehurst then employed David Graham as sales director (whom I had met while negotiating the Belstaff license) and another David who would continue as sales controller – so there were three Davids in the administration of the now-restructured Belstaff.

Unfortunately, Brocklehurst's ability as a managing director wasn't deemed to be as good as it could have been. I was back in Australia when the news came through the grapevine that Brocklehurst had resigned, and a couple of weeks later I was told that the new MD was a man called John Armstrong. I was about due to go to England, so I made it my business to look him up.

By this time our business with Belstaff was pretty significant to us and we were significant to Belstaff too because Driza-Bone had gradually become a more and more important part of their business and their own range of products had become less and less important. Belstaff was existing off Driza-Bone and not the other way around. They only continued to manufacture their own range of products for three years after that meeting, then they closed it down and concentrated on their Driza-Bone arm, which was doing very well.

'We Should Buy You Out!'

After attending management meetings for a couple of days, I was asked by Armstrong to travel to their factory in Silverton for discussions with their factory manager about how I wanted them to make Driza-Bone garments. This I did. I thought this was a bit odd, because we had well and truly gone through this before. I went out to the factory for three days, and on the third day Armstrong made me an offer as he was walking out with me after a long day of detailed examination.

As he and I were leaving the Belstaff premises, he said, "Frank, we should buy you out."

Those were his very words.

I wasn't interested at that time because we were in the throes of floating Armour-Driza-Bone Pty Ltd, which we looked like achieving in a very short time. We had the trademark registered in many countries and had formed Driza-Bone International Inc in the US.

Because our company had been closely audited by the underwriters, the prospectus requirements meant that Pat and I had a very clear idea of how much our company was worth at the time that Armstrong put his proposition.

Turn your mind back to convict days, to Peak Hill, Murwillumbah and the dances at the Petersham Town Hall. Turn your mind back to the awkward young man who asked Pat for a kiss under the mistletoe. To Nicholson Brothers & Lucas. To the heart attack that nearly killed me. To the Whitlam years which nearly send us to the wall, and to the rise of Australiana and my design of the coat. From this background, this early poverty, this struggle, the power of circumstance and the generosity of coincidence, I found myself talking in terms of millions of dollars, and everyone of the management of Belstaff was hanging on to my every word. To me, it was unbelievable.

The Pressure Builds

A lot of our profit was because we had used licensed manufacturers rather than making the capital investments ourselves, and it was not costing us very much to service these licensees. We also controlled the raw material, which we supplied to our licensed manufacturers, and so we made a three-way profit – on the supply of raw materials, on the licensee deal and from the retailers.

However, at that stage, our arrangements with Belstaff were so comprehensive that being a user of the same type of fabric, it became necessary for us to give them the name of our supplier, which enabled them to start placing their orders directly. This, coupled with the fact that we had been their manufacturers too, put them in a position of advantage over the others who depended on us.

In allowing them to source their own materials, we had our fears. Pat and I wanted control over the quality of the material that they were using. We knew from past experience that once a firm gets hold of a license they sometimes start cheapening it down. They want to cut costs and they don't have as deep a conscience about lessening the quality or streamlining production as the original manufacturer that knows and loves the product. We were not going to allow that to happen. We gave them the rights to purchase direct from our suppliers without paying us a commission, but we never allowed them to buy cheaper material or use material other than what was nominated by us.

Belstaff reciprocated by making us conform to their precise standards when we made their motorcycle-wear under license in Australia. We purchased material from their nominated source but the difference, however, was that we were only making maybe a couple of hundred per month of their garments whereas they could barely keep up with demand of Driza-Bone. By this time both products were being made from top-quality material. We all made very sure about that. Neither company would allow the other to downgrade.

The standards we set were so high that it sometimes created difficulties with suppliers. Both products had to be first class even though there's not always a lot of that high-standard fabric readily available. However, we managed to live up to the standards, and as time went along, we were able to buy in bigger quantities that guaranteed supply. Of course, once we got into larger quantities, economies of scale came into being and our prices could be sustained.

And so Pat and I had a lot to talk about in those days. Success is time consuming and maintaining the No 1 position can be exhausting even though it was important that we maintained control of Driza-Bone, as well as all our Armour Safety products. Ironically, the more we tried to maintain control of the company, the more the company started to control us. It's a bit like getting a tiger by the tail – you can't keep hanging on but you dare not let go. We were in that situation, going round and around in ever-increasing circles, wider and wider, bigger and bigger, and ultimately in fear that if we let go the tiger would come back and bite us for all sorts of reasons.

Success is like that. A company that is a household name can overtake your life because it is so complicated and so many people are involved – from the Australian Securities and Investments Commission (ASIC) to Princess Anne, and as far afield as the last retail shop before the Birdsville Track.

Selling Our Place In History

Pat, Stephen and I looked back on our lives and decided that we had achieved beyond our dreams, and if Belstaff were serious, then we would negotiate the sale of this business into which we had poured so much of our lives. In fact, the deal could have easily – and perhaps more logically – gone the other way around: we could have turned around and suggested to Belstaff that we buy them out.

In this, we only made one serious miscalculation, which was that in the process put into the selling, we thought that as 'Mr and Mrs Driza-Bone' the new owners would think that we meant something to the company. We thought that Stephen's massive contribution as Driza-Bone's No. 1 salesman would count for something to the new owners. We did not realise that when we sold our brand name, we were also selling everything – *everything!* – including a place in the history of the company.

The Belstaff owners – the Halstead Group (a listed public company in the UK) – had our figures confirmed by one of the largest accounting companies in the world. They also employed one of the largest financial companies in the world to be their underwriters. When the sale of Driza-Bone was finalised in February 1989, everything was agreed to as fair and reasonable by all parties.

Although Halstead's accountants privately told Pat that we had sold it too cheaply, it was still a very good price. Maybe we did sell a few hundred thousand on the cheap side but any rate, we sold it, and at the time we were very happy that we had achieved so much in our lifetime.

I'm not saying we were brilliant, but we were very good at doing what we did. Nobody would pay what Halstead paid unless Driza-Bone was a first-rate company.

Dark Horse

It was then that Belstaff/Halstead decided to bring John Maguire out from England, to take over my job in Eagleby, and that was the first time I properly met the man.

The financial director Steven Knight and MD John Armstrong were in Australia finalising the negotiations with Pat and me, when John Maguire literally took over my job, even though Pat and I were to be phased out over a five-week transfer period and not to make competitive garments for three years in Australia.

The first thing John Maguire did was exit Pat from the office that she had always occupied throughout all these years, and he stuck her into another office – a tiny one, and that was that. It was his company, he had the right to do this – but it was a rather blunt way of handling things.

I could see that it was time for the Fishers to let go, so I took John Maguire out to the factory floor, introduced him to all the staff and then he took over and I stepped aside. We turned up every morning for the five weeks and then on the sixth week we didn't. Of course, the money was great because everything was done by the book. We were paid all our money as well as a wage during the short period of time that Pat was to pass over her financial expertise to the new people. When we sold Driza-Bone we missed going to work – me more than Pat.

This was the saddest experience of my life: the realisation that as far as the new management was concerned, and despite all our efforts in the past, we were completely dispensable and probably – in their minds – a nuisance.

It was such a disappointment to us that the new management didn't spend a bit more time getting information from Pat, Stephen and I. But that was their decision. With dollars in our pockets, we had much to celebrate. Nevertheless, it's like a parent seeing their child leaving home. You train your child from birth to be as good a person as they can be, and we tried to do the same with the company. You like the assurance that they will make it into the future with the best chance.

After Pat and I left, Stephen continued working for them as marketing director of Armour Driza-Bone Pty Ltd. However, the new management didn't want *any* Fishers left in the company.

The End

I suppose that's what happens in a business takeover: new management takes over the old guard. However, I felt that I was very much part of the Driza-Bone production team and that we had devised innovative and more efficient ways of doing things, whereas Maguire wasn't a hands-on man.

Many times when he tried to get his ideas across, a staff member would call out, "Mr Fisher wouldn't do it that way" or "Mr Fisher and I used to work on it that way." He had the Devil's own job to take control because not only had the Fishers left a big impression on the company, but he was also an Englishman with a very strong Pommy accent. There was some sort of feeling about – "What would this bloke know about an outback riding coat?" – and for quite some time I received feedback from various staff members that I was not very popular with John Maguire.

Furthermore, by this time the Driza-Bone sale was on the TV News. The Channel 7 helicopter flew out to cover the story on the day we sold out. In fact, the helicopter landed on our own pad. The news focused on our past contribution as Australians, rather than the British connection, which must have unnerved the new management. In fact, over the period of three to four months leading up to the sale I was a big noise. I spoke at least 20 times about Driza-Bone and all that went with it – on radio, print media and TV stations.

Every few days there was a news story about Driza-Bone and every few days Frank Fisher, the former MD, said this, said that, said something. Everybody knew Frank, Pat and Stephen Fisher. I tried like mad to incorporate the message that the new management would continue to manufacture the product in Australia, and they have done so to date. Even though we had our differences, I was loyal to the new regime in my way – and ever loyal to the Driza-Bone coat, which I will always love.

Stephen Knight, financial director, and John Armstrong, managing director of Belstaff, publicly stated that the company was not going to close down production in Australia. That's what they stated, that they wanted to maintain the image of the Australian product. And the company is still situated in Eagleby in Queensland, about 60km from Brisbane's CBD where they still employ more than 150 people today.

But John Armstrong didn't last long. Belstaff was having problems in England and they closed down their manufacturing in Silverton and made their product offshore, on the continent.

That's the way the world goes. Production in high-cost countries suffers in labour-intensive industries and the rag trade is one. You can never make any money out of it if you use the traditional ways of doing things. We made money and Driza-Bone is still making money in Australia because we threw away some of the old traditional ways of making garments and came up with innovative methods of production, otherwise we wouldn't have survived either.

Legends, Lore And Many Lies

The whole mystique about Driza-Bone raincoats is they are made in the driest continent in the world. In buying Driza-Bone, Halstead/Belstaff bought the image of Australia. There were plenty of people who would not buy an English equivalent – not because the English product was second rate, but because it wasn't Australian. Ours was Australian and made in Australia, which was important. That's what Driza-Bone is all about.

After a time, in a weak attempt to claim a position in history, Halstead/Belstaff arranged for a book called *Legends, Lores and Lies* to be written about Driza-Bone. It is not much legend and mostly lies.

So many things in that book are patently wrong. It focuses on people who were never involved in Driza-Bone – some of whom didn't even *exist!* One man they talk about as being a 'legend' simply owned a hardware shop, another fellow was nothing more than a company salesman. These people never had any involvement with Driza-Bone during the period we owned it.

Furthermore, *Legends, Lores and Lies* doesn't even acknowledge Gordon Harman's contribution and his valiant attempt to keep the struggling business afloat over 25 difficult years.

It flicks through the Pickup family involvement, to which we owe so much.

And it doesn't even name me – I am just 'a gentleman from Australia' – and it says absolutely nothing about Pat or Stephen's massive contribution. The book reckons that Belstaff/Halstead decided to buy it from me after 'discussions on board a ship'.

I have no idea what they're talking about.

New Ideas

Maguire managed the company for some years. In certain ways he did a good job but he also made mistakes. One thing that worked well for him was that he replaced the principle of direct selling using the Driza-Bone sales team with a system of agents for all parts of Australia and he did a good job of that. Full marks. He had experience in using agents in England so he knew what he was doing. Furthermore, Maguire got very good agents who did a great job of selling Driza-Bone all around Australia, and overseas too.

However, Maguire did some other things that were quite unusual – and one of his decisions was to produce garments all year round, which is wrong, because Driza-Bone is a seasonal garment.

One of the things I could never understand was that the new Driza-Bone management didn't seem to value Stephen's years with the company. His knowledge and experience was most valuable as it was he more than anybody else who had the credibility when it came to selling Driza-Bone in country areas.

They threw Pat and me away, which may be all right, but to throw away Stephen's experience?

I never could understand it. Nor could he.

CHAPTER 18

Proving A Point

During the period when we owned Driza-Bone one of our licensed manufacturers was the New Zealand company Eidex, whose job was to manufacture the garments for New Zealand and the Pacific Islands. Eidex employed 40-50 people in their factory and at that stage they turned the factory across to making Driza-Bone garments, as well as their own range of padded jackets.

Eidex had actually been making oilskin garments as far back as the late 19th century. Some time in the past Eidex and Driza-Bone had developed close links with each other. The two companies had worked together in the past, helping each other with raw materials and in other ways. When we were trying to get fabrics during that difficult period after we first bought the Driza-Bone company, Gordon Harman contacted Eidex, as well as E Le Roy Ltd, and was able to get some oilskin fabrics off them too.

Eidex was named after a Swedish man called Oscar Eide who migrated to New Zealand during the 1800s and set up a manufacturing business. He became quite a large supplier of oilskin garments, tents and canvas goods. Well before the Driza-Bone sale, we had decided that we needed to get production done in New Zealand so we granted Eidex a license to manufacture Driza-Bone garments. Theirs was an ongoing license, not a yearly license.

We covered all the things that we both needed and we sealed the agreement with a handshake. We didn't need legal documents because we had mutual trust. Fred Westby, who was the MD of Eidex at the time, shook my hand and that's all we really needed.

However, we did put some specifics on paper which covered all the points that were important to our arrangement, so when Driza-Bone was sold to Belstaff they inherited the Eidex license agreement, although Halstead/Belstaff decided they didn't want to continue with Eidex.

Buying Eidex

By the late-1980s Eidex was part of a conglomerate that had been gradually winding down and as a company it was not longer profitable. I had often said to Pat, "When I get a bit older, I'll buy Eidex Ltd and add it to our group of companies."

Some six months after we sold Driza-Bone, we were told by Eidex Ltd that the company was going to be liquidated and closed down. Fred Westby, the managing director, asked if we were still interested in buying the company and I said, "Yes, I am."

Some time after, we purchased Eidex. I found out that the company sale had been publicised all over New Zealand and that it had been specifically offered to Driza-Bone, who had refused.

We bought Eidex because the cut-off with Driza-Bone was a bit sudden, particularly for me. I wanted to wind down my working life over a period of time until I was satisfied that the Driza-Bone company would continue to grow and do all the things I knew it was capable of doing. And so when Eidex became available, I bought it. We paid only a modest amount for it.

We acquired a lease of the factory in Wanganui, which is a town in the south-western part of the North Island, not far out of Wellington. We also acquired a group of factory workers who had some experience in making oilskin garments through having made Driza-Bone and Eidex products. We acquired sewing machines, cutting tables, a list of customers and a name – not a big name – but one that had a history.

We owned Eidex for four years, from 1989-1993, but John Maguire was not happy that I was once again involved in Driza-Bone which I now was because, through the Eidex purchase, I was now one of its licensees.

Maguire ordered me to stop making the Driza-Bone coats. I replied that I had a legitimate license and furthermore I was manufacturing it correctly – in an exemplary manner, in fact.

He disagreed, but our license was strong enough for him to be advised by his legal people that if they made a move on the New Zealand market while we still had this license, it would create complications. Despite this, Driza-Bone Australia decided to attack us by dumping Australian-made product on the New Zealand market, as he had threatened to do. He also decided to challenge our license in the courts, and after seeking legal advice we decided we would discontinue manufacturing Driza-Bone branded garments in New Zealand.

Once the matter got into the hands of the courts we knew we would have a six-to-12-month battle on our hands, so we revived the Eidex name and brought the oilskin garment back into the marketplace, having the Eidex patterns and a company history nearly as long as Driza-Bone. We then registered the name 'Eidex Kiwiwear' and we introduced the name 'Stockman Kiwiwear' into the range. I appointed a salesman who sold Eidex Kiwiwear Stockman coats throughout New Zealand, and he did a very good job of it. At this time I spent a lot of time in New Zealand though I never lived there permanently. We implemented exactly the same manufacturing, design and financial strategies over there as we had done with Driza-Bone in Australia.

At the end of 12 months I finally went to court in Wanganui – it was a case of Fisher v Driza-Bone, and I still ask myself how such a turn of events could ever have arisen – I who loved the coat!

We were told we were no longer licensed manufacturers and distributors for Driza-Bone garments in New Zealand, and we should cease using the name, which we had not used for nearly 12 months anyway.

During that period of time Maguire hadn't tried to take any of our New Zealand market. So while he was preparing to take us to court, having rebranded ourselves, we were madly out there selling Eidex Kiwiwear.

I was badly advised by my legal people and I lost the case. We paid $NZ22,000 for expenses and we went back into the marketplace as Eidex, which continued to grow.

Selling Eidex

In 1993 we sold the Eidex Ltd business and made another profit for the second time around. In the four years in which we owned it we got a handsome return on our money plus my wage which was quite a nice income.

The beautiful part about it as far as I was concerned was that it proved that what we had done with Driza-Bone wasn't mere luck. In that way we maintained our integrity and I proved my point. This was the most important thing as far as I was concerned. And I was very happy about that.

I introduced all the same principles of manufacturing which we had been proven with Driza-Bone and which Eidex knew about but hadn't fully taken on before I bought the company.

In the end, Eidex was bought by a Japanese company that operated in New Zealand as a car seat company. In those days Japanese cars were exported to New Zealand without the seats, and a New Zealand-based company made them. They made them for Datsuns, Toyotas and all the other imported Japanese cars. They made car seats to the exact specifications that they made them in Japan, which significantly reduced their import costs because they came in for duty at 5% while the car came in at 50%. Why bring in things you can make locally if you can beat the price?

In Japan, cars have to be sold for scrap after they are five years old. But five years is not very old, so in Japan a car has a limited life expectancy. At this time the Japanese started to export secondhand motor cars to New Zealand and when they started bringing in secondhand cars, the market for new cars dropped off, and with it, the market for car seat manufacturing.

So the company needed to look around for something else to do with their production facility. They had sewing machines for upholstery and a fully set-up factory, and when they saw our advertisement in a journal, they made contact with us and we sold Eidex to them.

CHAPTER 19

The Best Is Yet To Come!

I believe that all the key people with whom I have been associated throughout my life have taught me certain truths that made me the person that I am today. I think everybody can say the same, but in my case I happened to be in the right place at the right time as far as Driza-Bone was concerned. Furthermore, as the business grew, I happened to go to the right places at the right time and say the right things about Australia and Australiana.

Driza-Bone is certainly a part of Australia's history, and I will always be associated with the brand. But my involvement in the development of Australian Safety Standards has – in a way – brought me even more satisfaction than Driza-Bone.

I am probably the oldest person alive in Australia today who has made a career out of industrial safety. Now that you have read my book I hope you can sense the tremendous satisfaction that I have gained from working in this field, without which I never would have discovered and developed Driza-Bone.

Like many of my peers, who were there in the early days, I had a significant effect on safety procedures in Australia. We didn't *know* we were having an effect, we were just doing the job that needed to be done at the time.

As I have already told you, I was a member of the Chamber of Manufactures, the Safety Council of Australia, and the Standards Association of Australia. All these organisations made massive contributions towards improved safety standards – and I did my share of the things that needed to be done.

It was good fun too. I designed some of the proto-type goggles, but, more significantly, I was one of the official representatives to represent the Standards Association on industrial leather gloves. For a time, what I said became the standard by which all other leather gloves were judged.

Like my peers in the safety sector, I have obviously saved people a lot of injuries over the years so my involvement has given me a lot more satisfaction than I would have gained through working at other jobs. I have many examples of people who have since told me that I talked them into doing something that has saved lives or injuries. All this gave me a nice glow.

In those early days, industrial safety wasn't a highly profitable business because we had to spend a lot of time and energy convincing companies that they needed these things before they would buy them. Sometimes that was good; sometimes that was bad.

For example, I sold the idea of safety footwear to the British Leyland Corporation whose initial supplies was for 80 pairs of shoes bought from me (the firm had 5500 employees!). When they bought their next lot of shoes, they went direct to the manufacturer and cut me out. I lost their business even though I had worked my insides off to convince them to buy the safety footwear. Things like that have happened to me a number of times over the years, but ... that's business! After all, it's their job to keep costs down, whether I liked it or not.

As I said at the beginning, I am the sum total of all my ancestors, as well as the sum total of all my experiences. Those early years in western New South Wales gave me a grounding in the protective clothing needs of the working person, from how to protect fingers, eyes and toes, to how to keep the rain off your back when riding a horse.

Sydney Olympics 2000

Throughout my life, there have been many highlights, most of which I have shared in this book. However, one of the most emotional times for Pat and me was at the opening of the Sydney Olympics. We had been involved in previous Olympic Games, though never to the extent of what we saw on 15 September 2000 (the 15th again!).

Pru Acton had been a supplier to previous Olympic Games. When the 1988 Olympics were on in Seoul she and I co-opted with the maker of the uniforms, which is when I coined the expression 'Aussie Gold' because that was the first time that particular style, type and colour of material had ever been available in an oilskin garment.

By its very nature, oilskin tends to darken any colour. So we sourced a firm in Japan that made an acrylic fabric for picture theatre screens, and we made arrangements to buy our material from them in order to create those 'Aussie Gold' Driza-Bone coats. (The Seoul Driza-Bones were, therefore, not an oilskin coat in the true – or traditional – sense.)

When the Seoul Olympics were on, it was the busiest year we had ever had so Pat and I were too busy to attend the Games, and we certainly didn't get a chance to go to the expo in Brisbane that was on around the same time. However, at the expo our coats were given out to the representatives from each participating country as a gift from the Australian Government.

So when Pat heard that Gold Passes were being sold for the Sydney Olympics, Pat said to me, "For the first time in my life I want to attend the Games and witness an Opening Ceremony", so we bought one each.

The big night of the Opening Ceremony came.

Pat and I were seated in the stands.

Suddenly, this howl was heard throughout the arena – a huge groundswell of sound. I don't know where it came from, whether from the crowd or through the big speakers, this *roar!*

It built up and built up, and suddenly all these *Man From Snowy River* horseback riders raced out and started tearing around the arena, wearing our Driza-Bone coats!

Pat and I both had a good cry. It was a very emotional experience, it really was.

Today

Today, we live in Killara, Sydney, because it's handy to where most of our friends live. We have purchased many things that we only dreamed about owning during our working years. However, we have kept Pat's old piano because it was the very first thing Pat bought when she got her first job, and it is at Stephen's home in Lane Cove, Sydney.

Pat and I have always had a respect for old musical instruments. We have a *Beale* baby grand piano, which is of course an Australian instrument. We are also getting a 'box' piano restored. This 200-year old precursor of the piano deserves to be returned to the beautiful instrument that it was.

And I've still got my banjo mandolin.

After our active business life was over, Pat and I enjoyed a few overseas trips, which was a real pleasure because all our previous travels had a business agenda, so now we could relax and see the sights. We certainly like cruising on big boats, and during the past 10 years we have spent over a year of our total time on the ocean. We enjoy that type of life.

We did one trip that was three months long, on the *Royal Viking Sun* when she was one year old. Then, the following year we spent another four months on it. We have also travelled on the *QE2* and on a couple of other ocean liners, so we've done a fair bit of cruising in the past few years.

We have a two-bedroom unit on the Gold Coast at Runaway Bay, which Pat bought in the days when I was spending time in New Zealand getting the Eidex company going.

We've got a lot of friends on the Gold Coast and a lot of friends in Sydney, so we spend a bit of time betwixt and between.

Stephen

When we were negotiating the Driza-Bone sale, Stephen was a fully-fledged director of the company; his position was Marketing Director. So when we sold the business, we were led to believe that he was staying with Driza-Bone and that his future was assured. If I was buying a company and somebody like Stephen Fisher was available, I certainly would have held on to him and sent him back on the road to publicise the fact that it had new owners, while still retaining the old connection. I would have created a gradual change policy rather than a short, sharp break with the past.

Stephen was 'Mr Driza-Bone' in the country as far as the retailers were concerned, but the new management sent him out to sell an imported golf jacket from England, a product that was over-priced and not good enough. That was the approach they used to get rid of him. Stephen tried selling the stuff but he didn't have much success with it.

Stephen is not an aggressive type of person. After we left, he didn't feel like battling to retain his position in the company. In fact, he could have retained it because Maguire would have had to literally sack him to get rid of him. If he had stayed on he would probably have received a golden handshake before they could get rid of him, but he didn't want to make waves. He was only with them for some months after we left.

Stephen has since bought himself a coffee lounge in Runaway Bay. It was an upmarket coffee lounge, and again he was well accepted by the local people in the shopping centre. For a time Stephen's coffee shop was 'the place to go'. Quite a few well-known personalities used to frequent his place – among them, politician Ros Kelly and singer Olivia Newton-John.

Alas, it was an aging shopping centre and the powers-that-be decided to renovate it and increase its size. They gave Stephen an option to either make $500,000 worth of improvements or leave. He couldn't pour that amount of capital into his little business and still continue to make a profit, so he decided to sell his lease back to the centre. Since then the shop has gone through three different owners and gone broke each time. So Stephen clearly made the right decision.

He then bought a business importing reproduction furniture from Indonesia, which again was quite a successful little business. Unfortunately, Stephen has contracted cardiomyopathy, so I am sorry to report that he really hasn't been able to work for a few years.

Rolls Royce Club

I have always admired mechanical excellence and in 1990 I bought my first non-vintage Rolls Royce motor car. It was a 1986 model that I purchased to see if I liked it. I kept it for a year or so, sold it and in 1992 I bought a brand new Rolls Royce Silver Spirit II, which I've still got. (I've also got a little Honda 45cc motor scooter, which I ride around on. I call it the 'flying flea'.)

I bought the Rolls because I've always been a car fan and I wanted to buy just 'one more car' in my lifetime, so I thought I would purchase something that impressed me in every way. I joined the Rolls Royce Club five or six years ago and it gave me a lot of satisfaction being with a group of like-minded people, all of whom had an admiration for Rolls Royce motor cars. Members of the club get together regularly to talk about our cars and we sometimes go on rallies.

I have won several trophies over the years for the Best Model of that year or the Best Model of that type.

For quite a while I had an old 1939 Rolls Royce motor car. I spent a lot of time fixing it up. I rewired it, fixed up the doors and windows and did it all in my garage. I suppose all this started with the Dinky tricycle that I was given when I was four years old that I had to fix to get it going, so I did the same thing when I worked on the 1939 Rolls.

It's a simple motor car. It's not complicated. The wires were simple and went from A to B. I just pulled out the old wires and put new ones in their place. The 1939 Rolls had a carburetor, a distributor, a starter motor and all the usual things that you can get in and fix. So I have enjoyed that aspect of my life too.

Rotary

Thirty-four years have passed since I joined Rotary and as I said earlier in the book I am still enjoying it very much. I've been a member of the Ku-ring-gai Rotary Club for more than 20 years now. I am a past-President, and I have been awarded a Paul Harris Fellowship, which is a recognition for doing something special. I have been involved with a number of local projects. For example, I started *the Five Senses Walk* at the Wildflower Gardens in St Ives in 1984 – and I also instigated the path so that disabled people can enjoy the walk. So after all these years, I am still involved with the club's programs, including the Bowel Scan Program.

Another initiative was the Achievement Program, which is a management training program. It was started by the immediate past-President, Basil Lopis, but it actually began in my year as president. I get a little bit of satisfaction to have launched the program which has generated some $2-$3 million for Rotary in the past 16 years – and that's not bad for a Rotary Club!

I still get involved with various things in the club. In 2001, the Year of the Volunteer, we had a Carers' Night – 'carers' are people who 'care' for others but are seldom recognised for their contribution to society. So Rotary invited them to join us for a special dinner, where we publicly acknowledged their contribution to society.

We brought them up before the President individually and he presented them with a Certificate of Achievement and a small memento to take back to their 'caree'.

Live Above Your Income, And Achieve It

I can't imagine how lucky I was that a person like Pat was born into this world and that I was able to find her, even though in many ways we were geographically a million miles apart.

I lived in Campsie; Pat lived in Botany.

The likelihood of my meeting somebody from Botany was pretty small because in 1950 Campsie and Botany were 'a million miles' away from each other as far as Sydney was concerned. People who lived in Botany probably lived and died in Botany; people who live in Campsie probably did the same. So the chance of us meeting was very remote. But we did. And we became a great team.

I think the enjoyment we got out of the business was important too, and I think the actual business itself helped keep our marriage vital and alive. We were never bored with each other, or with our business. We owned the business and it owned us in many ways like all good businesses do.

Onwards and Upwards

Today, if you had to pay more than 10 per cent deposit you'd be screaming when you'd buy a house. In our day we had to pay 50% of the £4000 purchase price as deposit. But we said, "We will buy the house" and then we saved the deposit and we moved in.

We didn't have carpet on the floors; we just had polished floors for a while. We only owned a radio, a bed for Stephen, a bed for Pat and I, a table, four chairs and two-deck chairs, because we couldn't afford to buy furniture plus pay that 50% deposit. We were only there for a couple of years and then we decided we'd move on. We bought another house at Picnic Point. To do this, we had to sell the first house, and we made a small profit on it with which to pay the deposit on our next house.

Each step we made, we always bought a place that was dearer; we never went backwards. So when we bought a house, we always had that thought in mind – 'Buy above your income and then achieve that income'. We made a profit out of every house we bought – because they were good houses.

Furthermore, I always bought good motor cars. 'Good' as far as I was concerned meant practical and solid. They weren't mug lairish cars (although I must admit our Studebaker might have been a bit of mug lair's car). But apart from that one, we always bought cars that were right for the job.

Buyers do not buy from unsuccessful people. I believe if you're out on the road selling you should look the part. You should look as if you're successful – not too successful, but *successful*. A broken-down Holden does the same job as a Rolls Royce: they've both got four wheels and they both take you from A to B, but if you turn up to a selling job in a Rolls Royce you won't get the business because you look too successful. If you turn up in a 1952 Volkswagen you won't get the business either, because you don't look successful enough. You've got to position yourself somewhere in between, which I always tried to do.

We invested in ourselves. We were the best investment we knew of.

And the business never went backwards. The smallest profit we ever made over 12 months was $9000, in one very difficult year. It was one of those years when the country had a bad boom/bust period and we suffered like everybody else. So that's not too bad. We paid all our staff's wages, we paid ourselves a pittance, but we kept the business growing.

Enjoying Life

I guess to wrap up my life's history I would say: *I got involved in my life and I have enjoyed it.*

Generally I've enjoyed pretty good health, though I've had some nasty illnesses. I had a mastoid operation when I was nine, I had a quadruple by-pass when I was 65; I've had colitis and few other little problems like that over the years. I have Parkinson's now but it's very mild and it's not causing me any stress. I've had a few bad times, but I wouldn't change one thing that has ever happened to me. In their own way, even my sicknesses have had a positive side in changing my outlook and teaching me something I probably needed to learn about life.

I had my first heart attack when I was 39, which was not a good time to have a heart attack. The doctor said to me, "After you get well again, take up golf on Wednesday afternoons." Well ... I never did take up golf. But he also said, "Try to do your day's work in one day." He was quite a wise man and from then onwards Pat and I made a rule that we would try to work the same eight-hour day as the staff.

If we couldn't get our jobs done in time, it could only be because of one of two reasons: either (1) we had too much to do, or (2) we weren't efficient enough to do the job. So we had to make the appropriate adjustments to our personal habits and we always stayed closely within that framework.

We probably got through more work by disciplining our management of time, because we made a point of being able to get through what we needed to do.

Happiness

I have since met many famous people, mostly in connection with Driza-Bone. I have met Gough and Margaret Whitlam and there was no hostility on my part, because I believe that when he was PM Whitlam was doing the job he had to do. He was given that job by his people, and he just did it. The downside was that he wrecked our Armour Safety business, and the upside was he forced me to think outside the square – which led to the purchase of the run down Driza-Bone company ... and the rest is history.

Of course, I've still got my Driza-Bone coat.

It's hanging up in the wardrobe at home.

We were very successful at doing what we did and we are very proud of ourselves knowing that we did a good job. I'm not skiting; the results are there.

I don't think I can say much more about my life than that.